ATS CODING ACADEMY

PYTHON

PYTHON PROGRAMMING FOR BEGINNERS

An Easy and Step-by-Step Guide for Absolute Beginners

Learn Programming Fast

BRIAN JENKINS

How to contact us

If you find editing issues or any other issues in this book, please immediately notify our customer service by email at:

customer_service@atscodingacademy.com

If you have any questions or suggestions, you can also contact the author directly at:

brian.jenkinks@atscodingacademy.com

or

brian.jenkins.m@outlook.com

ATS Coding Academy provide you high-quality books for your technical learning in Computer Science and Programming subjects.

Thank you so much for buying this book.

© Copyright 2018 by Brian Jenkins.

All rights reserved.

First Printing, 2017

Edited by Edgar Dobson

E-book Converted and Covered by Warne's House

Published by CreateSpace Publishing

ISBN-13: 978-1719103282

ISBN-10: 1719103283

The contents of this book may not be reproduced, duplicated or transmitted without the direct written permission of the author.

Under no circumstances will any legal responsibility or blame be held against the publisher for any reparation, damages, or monetary loss due to the information herein, either directly or indirectly.

Legal Notice:

You cannot amend, distribute, sell, use, quote or paraphrase any part or the content within this book without the consent of the author.

Disclaimer Notice:

Please note the information contained within this document is for educational and entertainment purposes only. No warranties of any kind are expressed or implied. Readers acknowledge that the author is not engaging in the rendering of legal, financial, medical or professional advice. Please consult a licensed professional before attempting any techniques outlined in this book.

By reading this document, the reader agrees that under no circumstances is the author responsible for any losses, direct or indirect, which are incurred as a result of the use of information contained within this document, including, but not limited to, errors, omissions, or inaccuracies.

Introduction

"Talk is cheap. Show me the code."

Linus Torvalds

Python is an object-oriented and multi-purpose programming language. Python can be used to develop different types of applications ranging from web applications, desktop applications and even game applications. The language is popular for its syntax which is easy for one to grasp. This has made it one of the best coding languages for absolute beginners to computer programming. This is the reason as to why Python is taught in junior and middle-aged students in schools. The language was developed with the goal of making programing easy for understanding.

Python is a cross-platform programming language. It can be used on various systems including Windows, Unix, Mac etc. It also comes with numerous modules that are cross-platform. However, Python has maintained a uniform user interface. Python supports integration with different database management systems (DBMSs). This means that you can access your data from a database with Python. This book explores every aspect of Python programming language. It will help you in preparing a solid computer programming foundation and learning any other coding language will be easy to you.

Book Objectives

The author wrote this book with the goal of helping the readers learn every aspect of Python programming.

This book will help you:

- Know more about computer programming and how to get started with Python programming language.
- Understand the various features of Python programming language and appreciate its power.
- Transition from a programming beginner to an expert.

Target Users

The book designed for a variety of target audiences. The most suitable users would include:

- Newbies in computer programming and Python Programming
- Professionals in computer programming and software applications development
- Professors, lecturers or tutors who are looking to find better ways to explain the content to their students in the simplest and easiest way
- Students and academicians, especially those focusing on computer programming and software development

Is this book for me?

If you want to learn computer programming with Python, this book is for you. Experience in computer programming is not required. If this is the first time for you to hear about computer programming, this book is the best for you.

To my wife Chelsea and my daughter Britany

You are the happiness of my days, and you are my truest love!

Author Biography

Dr. Brian Jenkins received the B.S. degree in computer science from EPITAE, in France, the masters and a Ph.D. degree in computer science from HESSIO in Switzerland. He worked with Seagate Research, Pittsburgh, PA, and with Tyco Communications Laboratories, Eatontown, NJ, conducting coding research for magnetic recording systems and long-haul fiber optic communication systems.

With two decades experience teaching programming to newcomers, and one of the most talented IT specialists of his generation, Brian works as a Research Software Specialist and occasional bioinformatician.

Brian has been in the software field for over 20 years and explored implementations of the Prolog language, and over his career has worked as a professional software developer on compilers, programming tools, scripting systems, and assorted client/server and business applications.

In addition, Brian worked as a research and teaching assistant at the Chair of Information and Coding Theory (ICT) towards his Ph.D. degree at the Faculty of Engineering.

Since 2015 Brian resides in Geneva, in Switzerland with his wife, Chelsea, and daughter, Britany. He is working on series of books in programming and computer science..

Table of Contents

Introduction 7

Chapter 1- Getting Started with Python

What is Python? 19

Installing Python 20

    Installation on Windows 21

    Installation on Linux 21

    Installation on Mac OS 21

Running Programs 22

    Interactive Interpreter 22

    Script from Command Line 23

    Python IDE (Integrated Development Environment) 24

Chapter 2- Basic Python Syntax

Indentation 26

Quotes 27

User Input 28

Chapter 3- Python Variables

Multiple Variable Assignment 31

Chapter 4- Python Data Types

Python Numbers 34

Python Strings 37

Python Lists 38

Python Tuples 39

Python Dictionaries 41

Datatype Conversion 43

Chapter 5- Control Statements

If Statement 46

If-Else Statement 48

If Elif Else Statement 50

Nested If 53

Chapter 6- Python Functions

Function Parameters 57

Function Parameter Defaults 61

Chapter 7- Python Loops

For Loop 64

"RANGE()" function 64

While Loop 68

Loop Control 71

Break Statement 71

Continue Statement 73

Pass Statement 74

Chapter 8- Python Classes and Objects

Class Definition 76

Built-in Attributes 79

Garbage Collection 81

Inheritance 83

Multiple Inheritance 86

Python Constructors 88

Overriding Class Methods 88

Operator Overloading..90

Chapter 9- Exception Handling

Raising Exceptions 100

Exception Objects..101

Custom Exception Class 102

Chapter 10- Python Modules

Locating Modules 110

Namespaces and Scope 111

dir() Function 114

locals () and global() Functions 114

reload () Function 115

Chapter 11- File Handling

open() Function 116

close() Method 119

write() Method 120

read () Method...121

File Positions 122

rename() Method 123

remove() Method 124

mkdir() Method...125

chdir() Method 126

getcwd() Method 126

rmdir() Method 127

Chapter 12- Tkinter

TKinter Buttons 130

TKinter MenuButtons 131

Canvas 136

Slider 138

TKinter Label 140

TKinter Checkbutton 141

TKinter Radiobutton 143

Chapter 13- Python Operators

Arithmetic Operators 148

Comparison Operators 150

Assignment Operators 152

Membership Operators 154

Identity Operators 155

Chapter 14- Accessing MySQL Databases

Creating a Table 161

Inserting Data 162

Conclusion

Chapter 1- Getting Started with Python

What is Python?

Python is a programming/coding language. It's one of the programming languages that are interpreted rather than compiled. This means the Python Interpreter works or operates on Python programs to give the user the results. The Python Interpreter works in a line-by-line manner. With Python, one can do a lot. Python has been used for development of apps that span a wide of fields, from the most basic apps to the most complex ones. Python can be used for development of the basic desktop computer applications. It is also a good coding language for web development. Websites developed with Python are known for the level of security and protection they provide, making them safe and secure from hackers and other malicious users. Python is well applicable in the field of game development. It has been used for development of basic and complex computer games. Python is currently the best programming language for use in data science and machine learning. It has libraries that are best suitable for use in data analysis, making it suitable for use in this field. A good example of such a library is scikit-learn (sklearn) which has proved to be the best for use in data science and machine learning.

Python is well known for its easy-to-use syntax. It was written with the goal of making coding easy. This has made it easy the best language even for beginners. Its semantics are also easy, making it easy for one to understand Python codes. The language has received a lot of changes and improvements, especially after the introduction of Python 3. Previously, we had Python 2.7 which had gained much stability. Python 3 brought in new libraries, functions and other features, and some of the language constructs where changed significantly. The names of some Python libraries were also changed, especially when it comes to case.

Installing Python

To code in Python, you must have the Python Interpreter installed in your computer. You must also have a text editor in which you will be writing and saving your Python codes. The good thing with Python is that it can run on various platforms like Windows, Linux and Mac OS. Most of the current versions of these operating systems come installed with Python. You can check whether Python has been installed on your operating system by running this command on the terminal or operating system console:

Python

Type the above command on the terminal of your operating system then hit the Enter/Return key. The command should return the version of Python installed on your system. If Python is not installed, you will be informed that the command is not recognized, hence you have to install Python.

Installation on Windows

To install Python on Windows, download Python from its official website then double click the downloaded setup package to launch the installation. You can download the package by clicking this link:

https://www.python.org/downloads/windows/

It will be good for you to download and install the latest package of Python as you will be able to enjoy using the latest Python packages. Currently, we have Python 3.6 being the latest release of Python. After downloading the package, double click it and you will be guided through on-screen instructions on how to install Python on your Windows OS.

Installation on Linux

In Linux, there are a number of package managers that can be used for installation of Python in various Linux distributions. For example, if you are using Ubuntu Linux, run this command to install Python:

$ sudo apt-get install python3-minimal

Python will be installed on your system. However, most of the latest versions of various Linux distributions come installed with Python. Just run the "python" command. If you get a Python version as the return, then Python has been installed on your system. If not, go ahead and install Python.

Installation on Mac OS

To install Python in Mac OS, you must first download the package. You can find it by opening the following link on your web browser:

https://www.python.org/downloads/mac-osx/

After the setup has been downloaded, double click it to launch the installation. You will be presented with on screen instructions that will guide through the installation process. Lastly, you will have Python running on your Mac OS system.

Running Programs

One can run Python programs in two main ways:

Interactive interpreter

Script from command line

Interactive Interpreter

Python comes with a command line which is commonly referred to as the interactive interpreter. You can write your Python code directly on this interpreter and press the enter key. You will instant results. If you are on Linux, you only have to open the Linux terminal then type "python". Hit the enter key and you will be presented with the Python interpreter with the >>> symbol. To access the interactive Python interpreter on Windows, click Start -> All programs, then identify "Python …" from the list of programs. In my case, I find "Python 3.5" as I have installed Python 3.5. Expand this option and click "Python …". In my case, I click "Python 3.5(64-bit)" and I get the interactive Python interpreter.

Here, you can write and run your Python scripts directly. To write the "Hello" example, type the following on the interpreter terminal:

print("Hello")

Hit the enter/return key and the text "Hello" will be printed on the interpreter:

Script from Command Line

This method involves writing Python programs in a file, then invoking the Python interpreter to work on the file. Files with Python could should be saved with a .py extension. This is a designation to signify that it is a Python file. For example, script.py, myscript.py etc. After writing your code in the file and saving with the name "mycode.py", you can open the operating system command line and invoke the Python interpreter to work on the file. For example, you can run this command on the command line to execute the code on the file mycode.py:

python mycode.py

The Python interpreter will work on the file and print the results on the terminal.

Python IDE (Integrated Development Environment)

If you have a GUI (Graphical User Interface) application capable of supporting Python, you can run the Python on a GUI environment. The following are the Python IDEs for the various operating systems:

- UNIX- IDLE
- Windows- PythonWin
- Macintosh- this comes along with IDLE IDE, downloadable from the official website as MacBinary or BinHex'd files.

Chapter 2- Basic Python Syntax

As stated earlier, Python was written with the goal of making programming easy. Its syntax is closely related to the one used in popular coding languages like C and Java. In our previous example, we wrote the following statement:

print("Hello")

What we are doing is instructing Python interpreter to print Hello on the terminal. This has been achieved by calling the print() function. Functions are predefined, so the Python interpreter will understand what you mean when you call the function. Python functions are written using parenthesis (), which is a designation to mean that you are writing a function.

Indentation

Most programming languages rely on curly braces {} to group blocks of statements that are related or the ones that are to perform a unit task. This is not the case with Python. Python relies on indentation to create blocks of statements. Statements in same block must have similar indentation. Example:

if True:

   print ("Condition True")

else:

   print ("Condition False")

The "if" and "else" statements have same level of indentation. Example 2:

if True:

   print ("Condition True")

   print ("Printing True ")

else:

   print "(Condition False")

 print ("Printing False")

In the above code, we will get an error after running the script. The last two print statements belong to similar block but they have not been indented to same level. They should be indented as in the first two print statements.

Quotes

Python accepts single, double and triple quotes. They help in enclosing string literals. In our previous statement:

print("Hello")

We have opened the string to be printed with double quotes and closed the string with double quotes. If you open with a particular type of quote, you must use it to close the string, otherwise, an error will be generated. We could also have used single or triple quotes to enclose the string and the result would have been the same. Example:

Using single quotes:

print('Hello')

Using triple quotes:

print('''Hello''')

However, these were not introduced to be used this way. Triple quotes should be used when there is a need to span a particular string across a number of lines. Single quotes should be used to quote a word, while the double quotes should be used to quote a sentence. Example:

word = 'hello'

sentence = "It's a sentence."

paragraph = """It's a paragraph

in Python with multiple lines"""

User Input

When writing your program or creating an application, you may require the users to enter an input such as their username and other details. Python provides the input() function that helps you get and process input from users. Other than entering input, you may require the users to perform an action so that they may go to the next step. For example, you may need them to press the enter key on the keyboard to be taken to next step. Example:

input("\n\n Press Enter key to Leave.")

Just type above statement on the interactive Python interpreter then hit the Enter key on the keyboard. You will be prompted to press the Enter key:

```
>>> input("\n\n Press Enter key to Leave.")

 Press Enter key to Leave.
```

The program waits for an action from the user to proceed to next step. Notice the use of \n\n which are characters to create a new line. To create one line, we use a single one, that is, \n. In this case, two blank lines will be created. That is how Python input() function works.

Chapter 3- Python Variables

Python variables preserve a location in memory that can be used for storage of values. Once a variable is created, some memory space is reserved for it. Variables are of different types, and the type used to declare the variables determines the amount of storage space assigned to the variables as well as the value that can be stored in that variable. The equal sign (=) is used for assignment of a value to a variable. When a value has been assigned to a variable, that variable will be declared automatically.

Example:

#!/usr/bin/python3

age = 26   # Integer variable and value

height   = 17.1   # Floating point variable and value

name   = "Nicholas"   # String variable and value

print (age)

print (height)

print (name)

After running the above program, you will get the following result:

```
26
17.1
Nicholas
```

We declared three variables namely age, height and name. The three were also assigned values. We have then used the print function to access the values of these functions and print them on the terminal. Note that the variables have not been enclosed within quotes in the print statement. This is because we are accessing variables that have been defined already.

Multiple Variable Assignment

In Python, a single value can be assigned to a number of variables at once. Example:

a = b = c = 2

In the above example, the value 2 has been assigned to three different variables namely a, b and c. This means each of these variables has a value of 2. The three will also be kept in a single location.

It is also possible for you to assign multiple objects to different variables. Example:

a, b, c = 2, 3, "nicholas"

In the above example, the values will be assigned to the variables according to their order, Variable a will be assigned a value of 2, b a 3 and c "nicholas".

You can run this program to access the values of individual variables given above:

#!/usr/bin/python3

a, b, c = 2, 3, "nicholas"

print(a)

print (b)

print (c)

This will give the result shown below:

```
2
3
nicholas
```

It is very clear that the values were assigned to the variables based on their order.

Chapter 4- Python Data Types

Python supports different data types. Each variable should belong to one of the data types supported in Python. The data type determines the value that can be assigned to a variable, the type of operation that may be applied to the variable as well as amount of space assigned to the variable. Let us discuss different data types supported in Python:

Python Numbers

These data types help in storage of numeric values. . The creation of number objects in Python is done after we have assigned a value to them. Consider the example given below:

total = 55

age= 26

You are familiar with this as we had discussed it earlier. Also, it is possible for you to delete a reference to a particular number variable. This can be done by use of the del statement. This statement takes the following syntax:

del variable1[,variable2[,variable3[....,variableN]]]

The statement can be used for deletion of a single or multiple variables. This is shown below:

del total

del total, age

In the first statement, we are deleting a single variable while in the second statement, we are deleting two variables. If the variables to be deleted are more than two, separate them by use of a comma and they will be deleted.

In Python, there are four numerical values which are supported:

- Int
- Float
- complex

In Python3, all integers are represented in the form of long integers.

The Python integer literals belong to the int class. Example:

Run the following statements consecutively on the Python interactive interpreter:

x=10

x

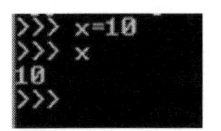

```
>>> x=10
>>> x
10
>>>
```

The float is used for storing numeric values with a decimal point. Example:

x=10.345

x

You can run it on the Python interactive interpreter and you will observe the following

```
>>> x=10.345
>>> x
10.345
>>>
```

If you are performing an operation with one of the operands being a float and the other being an integer, the result will be a float. Example:

5 * 1.5

```
>>> 5 * 1.5
7.5
>>>
```

As shown above, the result of the operation is 7.5 which is a float.

Complex numbers are made of real and imaginary parts, with the imaginary part being denoted using a j. They can be defined as follows:

x = 4 + 5j

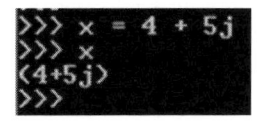

```
>>> x = 4 + 5j
>>> x
(4+5j)
>>>
```

In above example, 4 is the real part while 5 is the imaginary part.

Python with a function named type() that can be used for determination of the type of a variable. You only have to pass the name of the variable inside that function as the argument and its type will be printed. Example:

x=10

type(x)

```
>>> x=10
>>> type(x)
<class 'int'>
>>>
```

The variable x is of int class as shown above. You can try it for other variable types as shown below:

name='nicholas'

type(name)

```
>>> name='nicholas'
>>> type(name)
<class 'str'>
>>>
```

The variable is of the string class as shown above.

Python Strings

Python strings are series of characters enclosed within quotes. Use any type of quotes to enclose Python strings, that is, either single, double or triple quotes. To access string elements, we use the slice operator. String characters begin at index 0, meaning that the first character string is at index 0. This is good when you need to access string characters. To concatenate strings in Python, we use + operator, the asterisk 9*) is used for repetition. Example:

#!/usr/bin/python3

thanks = 'Thank You'

print (thanks)   # to print the complete string

print (thanks[0])   # to print the first character of string

print (thanks[2:7])   # to print the 3rd to the 7th character of string

print (thanks[4:])   # to print from the 5th character of string

print (thanks * 2)   # to print the string two times

print (thanks + "\tAgain!") # to print a concatenated string

The program prints the following once executed:

```
Thank You
T
ank Y
k You
Thank YouThank You
Thank You      Again!
```

Notice that we have text beginning with # symbol. The symbol denotes beginning of a comment. The Python print will not act on the text from the symbol to the end of the line. Comments are meant at enhancing the readability of code by giving explanation. We defined a string named *thanks* with the value *Thank You*. The *print (thanks[0])* statement helps us access the first character of the string, hence it prints T. You also notice that the space between the two words is counted as a character.

Python Lists

Lists consist of items enclosed within square brackets ([]) and the items are separated using commas (,). They are similar to the C arrays. Although all array elements must belong to similar type, lists supports the storage of items belonging to different types in a single list.

We use the slice operator ([ ] and [:]) for accessing the elements of a list. The indices start from 0 and end at -1. Also, the plus symbol (+) represents the concatenation operator while the asterisk (*) represents the repetition operator. Example:

#!/usr/bin/python3

listA = ['john', 3356 , 8.90, 'sister', 34.21]

listB = [120, 'sister']

print listA   # will print the complete list

print listA[0]   # will print the first element of the list

print listA[1:3] # will print the elements starting from the 2nd till 3rd

print listA[2:]   # will print the elements starting from the 3rd element

print listB * 2   # will print the list two times

print listA + listB # will print a concatenated lists

There is no much difference in what is happening in the above code compared to the previous one for strings. When executed, the program outputs:

```
['john', 3356, 8.9, 'sister', 34.21]
john
[3356, 8.9]
[8.9, 'sister', 34.21]
[120, 'sister', 120, 'sister']
['john', 3356, 8.9, 'sister', 34.21, 120, 'sister']
```

In the statement "print listA", we print the contents of listA. Note that each element is treated to be at its own index as a whole, for example, element 'john' is treated as a single element of a list at index 0.

Python Tuples

Python tuples are similar to lists with the difference being after creating a tuple, you cannot add, delete or change the tuple elements. Tuple elements should be enclosed within parenthesis (). Example:

#!/usr/bin/python3

t1 = () # creating an empty tuple, that is, no data

t2 = (22,34,55)

t3 = tuple([10,23,78,110,89]) # creating a tuple from an array

t4 = tuple("xyz") # creating tuple from a string

print t1

print t2

print t3

print t4

The values of the 4 tuples will be printed:

```
()
(22, 34, 55)
(10, 23, 78, 110, 89)
('x', 'y', 'z')
```

There are a number of functions that can be applied on tuples. Example:

#!/usr/bin/python3

t1 = (23, 11, 35, 19, 98)

print("The minimum element in the tuple is", min(t1))

print("The sum of tuple elements is", sum(t1))

print("The maximum element in the tuple is", max(t1))

print("The tuple has a length of", len(t1))

When executed, it gives this result:

```
The minimum element in the tuple is 11
The sum of tuple elements is 186
The maximum element in the tuple is 98
The tuple has a length of 5
```

First, we called the *min()* function which returns the smallest element in the tuple. We then called the *sum()* function which returned the total sum of tuple elements. The *max()* function returned the maximum element in the tuple. The *len()* function counted all elements in the tuple and returned their number.

You can use the slice operator to access some of the tuple elements, not all. Example:

#!/usr/bin/python3

t = (23, 26, 46, 59, 64)

print(t[0:2])

When executed, it prints:

```
(23, 26)
```

We have used the slice operator to access elements from index 0 to index 2 in the tuple. Note that tuple elements begin at index 0.

Python Dictionaries

Python dictionaries are used for storage of key-value pairs. With dictionaries, you can use a key to retrieve, remove, add or modify values. Dictionaries are also mutable, meaning you can't their values once declared.

To create dictionaries, we use curly braces. Every dictionary item has a key, then followed by colon, then a value. The items are separated using a comma (,). Example:

#!/usr/bin/python3

classmates = {

'john' : '234-221-323',

'alice' : '364-32-141'

}

We have created a dictionary named classmates with two items. Note that the key must be of a type that is hashable, but you may use any value. Each dictionary key must be unique. I first element, john is the key followed by the value. In second element, alice is the element. To access dictionary elements, use the dictionary name and the key. Example:

#!/usr/bin/python3

classmates = {

'john' : '234-221-323',

'alice' : '364-32-141'

}

print("The number for john is", classmates['john'])

print("The number for alice is", classmates['alice'])

The last two statements help us access the dictionary values. It prints:

```
The number for john is 234-221-323
The number for alice is 364-32-141
```

To know the dictionary length, run the len() function as follows:

len(classmates)

The above will return 2 as the dictionary has only two elements.

Datatype Conversion

Python allows you to convert data from one type to another. The process of converting from one datatype to another is known as *typecasting*.

If you need to convert your *int* datatype into *a float*, you call the *float()* function. Example:

#!/usr/bin/python3

height=20

print("The value of height in int is", height)

print("The value of height in float is", float(height))

In above example, height has been initialized to 20. We have called the float() function and passed height to it as the parameter. The integer value, that is, 20 has been converted into a float value, that is, 20.0. The program prints the following:

```
The value of height in int is 20
The value of height in float is 20.0
```

To convert a float into an int, you call the int() function. Example:

#!/usr/bin/python3

height=20.0

print("The value of height in float is", height)

print("The value of height in int is", int(height))

The program prints the following:

```
The value of height in float is 20.0
The value of height in int is 20
```

We have called the *int()* function and passed the parameter *height* to it. It has converted 20.0 to 20, which is a float to an integer conversion.

If you need to convert a number to a string, you call the *str()* function. The number will then be converted into a string. Example:

#!/usr/bin/python3

num=20

print("The value of num in int is", num)

print("The value of num in string is", str(num))

The program outputs:

```
The value of num in int is 20
The value of num in string is 20
```

Although the value is the same, it is treated differently by Python interpreter. The conversion of a float to a string can also be done similarly.

Chapter 5- Control Statements

Sometimes, you may need to run certain statements based on conditions. The goal in control statements is to evaluate an expression or expressions, then determine the action to perform depending on whether the expression is TRUE or FALSE. There are numerous control statements supported in Python:

If Statement

With this statement, the body of the code is only executed if the condition is true. If false, the statements after If block will be executed. It is a basic conditional statement in Python. Example:

```
#!/usr/bin/python3

ax = 7

bx = 13

if ax > bx:

    print('ax is greater than bx')
```

The above code prints nothing. We defined variables *ax* and *bx*. We then compare their values to check whether ax is greater than bx. This is false, hence nothing happens. The > is "greater than" sign. Let us change it to >, that is, less than sign:

```
#!/usr/bin/python3


ax = 7
```

bx = 1

if ax < bx:

   print('ax is greater than bx')

This prints the following:

```
ax is greater than bx
```

The condition/expression was true, hence the code below the If expression is executed. Sometimes, you may need to have the program do something even if the condition is false. This can be done with indentation in the code. Example:

#!/usr/bin/python3

ax = 10

if ax < 5:

   print ("ax is less than 5")

   print (ax)

if ax > 15:

   print ("ax is greater than 15")

   print (ax)

print ("No condition is True!")

In the above code, the last *print()* statement is at the same level as the two Ifs. This means even any of the two is true, this statement will not be executed. However, the statement will be executed if both Ifs are false.

Running the program outputs this:

```
No condition is True!
```

The last *print()* statement as executed as shown in result above.

If-Else Statement

This statement helps us specify a statement to execute in case the If expression is false. If the expression is true, the If block is executed. If the expression is false, the Else block will run. The two blocks cannot run at the same time. It's only one of the that can run. It is an advanced If statement.

Example:

#!/usr/bin/python3

ax = 10

bx = 7

if ax > 30:

   print('ax is greater than 30')

else:

   print('ax isnt greater than 30')

The code will give this result once executed:

```
ax isnt greater than 30
```

The value of variable *ax* is 30. The expression *if ax* > 30: evaluates into a false. As a result, the statement below *If*, that is, the first *print()* statement isn't executed. The else part, which is always executed when the If expression is false will be executed, that is, the *print()* statement below the *else* part.

Suppose we had this:

```
#!/usr/bin/python3

ax = 10

bx = 7

if ax < 30:

    print('ax is less than 30')

else:

    print('ax is greater than 30')
```

This will give this once executed:

```
ax is less than 30
```

In the above case, the print() statement within the If block was executed. The reason is because the If expression as true. Another example:

```
#!/usr/bin/python3

ax = 35

if ax % 2 ==0:

    print("It is eve")
```

else:

  print("It is odd")

The code outputs:

```
It is odd
```

The If expression was false, so the else part was executed.

If Elif Else Statement

This statement helps us test numerous conditions. The block of statements under the *elif* statement that evaluates to true is executed immediately. You must begin with *If* statement, followed by *elif* statements that you need then lastly the *else* statement, which must only be one. Example:

#!/usr/bin/python3

ax = 6

bx = 9

bz = 11

if ax > bx:

  print('ax is greater than bx')

elif ax < bz:

  print('ax is less than bz')

else:

  print('The else part ran')

The code outputs the following:

```
ax is less than bz
```

We have three variables namely *ax*, *bx* and *bz*. The first expression for *If* statement is to check whether ax is greater than bx, which is false. The *elif* expression checks whether *ax* is less than *bx*, which is true. The *print()* statement below this was executed. Suppose we had this:

#!/usr/bin/python3

ax = 6

bx = 9

bz = 11

if ax > bx:

   print('ax is greater than bx')

elif ax > bz:

   print('ax is less than bz')

else:

   print('The else part ran')

The code will output:

```
The else part ran
```

In the above case, both the *If* and *elif* expressions are false, hence the *else* part was executed. Another example:

#!/usr/bin/python3

```python
day = "friday"
if day == "monday":
    print("Day is monday")
elif day == "tuesday":
    print("Day is tuesday")
elif day == "wednesday":
    print("Day is wednesday")
elif day == "thursday":
    print("Day is thursday")
elif day == "friday":
    print("Day is friday")
elif day == "saturday":
    print("Day is saturday")
elif day == "sunday":
    print("Day is sunday")
else:
    print("Day is unkown")
```

The value of *day* if *friday*. We have used multiple *elif* expressions to check for its value. The *elif* expression for *friday* will evaluate to true, hence its *print()* statement will be executed.

Nested If

An *If* statement can be written inside another *If* statement. That is how we get nested *If*. Example:

```
#!/usr/bin/python3

day = "holiday"

balance = 110000

if day == "holiday":
  if balance > 70000:
    print("Go for outing")
  else:
    print("Stay indoors")
else:
  print("Go to work")
```

We have two variables *day* and *balance*. The code gives the following result:

```
Go for outing
```

The first *if* expression is true as it's holiday. The second *if* expression is also true since balance is greater than 70000. The *print()* statement below that expression is executed. The execution of the program stops there. Suppose the balance is less than 70000 as shown below:

```
#!/usr/bin/python3
```

```
day = "holiday"

balance = 50000

if day == "holiday":

  if balance > 70000:

    print("Go for outing")

  else:

    print("Stay indoors")

else:

  print("Go to work")
```

The value of *balance* is 50000. The first *if* expression is true, but the second one is false. The nested *else* part is executed. We get this result from the code:

```
Stay indoors
```

Note that the nested part will only be executed if and only if the first *if* expression is true. If the first *if* is false, then the un-nested *else* part will run. Example:

```
#!/usr/bin/python3

day = "workday"

balance = 50000

if day == "holiday":

  if balance > 70000:
```

        print("Go for outing")
    else:
        print("Stay indoors")
else:
    print("Go to work")

The value for *day* is *workday*. The first *if* expression testing whether it's a holiday is false, hence the Python interpreter will move to execute the unnested *else* part and skip the entire nested part. The code gives this result:

```
Go to work
```

Chapter 6- Python Functions

Python functions are a good way of organizing the structure of our code. The functions can be used for grouping sections of code that are related. The work of functions in any programming language is improve the modularity of code and make it possible to reuse code.

Python comes with many in-built functions. A good example of such a function is the "print()" function which we use for displaying the contents on the screen. Despite this, it is possible for us to create our own functions in Python. Such functions are referred to as the "user-defined functions".

To define a function, we use the "def" keyword which is then followed by the name of the function, and then the parenthesis (()).

The parameters or the input arguments have to be placed inside the parenthesis. The parameters can also be defined within parenthesis. The function has a body or the code block and this must begin with a colon (:) and it has to be indented. It is good for you to note that the default setting is that the arguments have a positional behavior. This means that they should be passed while following the order in which you defined them. Example:

#!/usr/bin/python3

def functionExample():

    print('The function code to run')

    bz = 10 + 23

    print(bz)

We have defined a function named *functionExample*. The parameters of a function are like the variables for the function. The parameters are usually added inside the parenthesis, but our above function has no parameters. When you run above code, nothing will happen since we simply defined the function and specified what it should do. The function can be called as shown below:

```
#!/usr/bin/python3

def functionExample():
    print('The function code to run')
    bz = 10 + 23

functionExample()
```

It will print this:

```
The function code to run
```

That is how we can have a basic Python function.

Function Parameters

You can dynamically define arguments for a function. Example:

```
#!/usr/bin/python3

def additionFunction(n1,n2):
    result = n1 + n2
    print('The first number is', n1)
```

print('The second number is', n2)

print("The sum is", result)

additionFunction(10,5)

The code returns the following result:

```
The first number is 10
The second number is 5
The sum is 15
```

We defined a function named *addFunction*. The function takes two parameters namely *n1* and *n2*. We have another variable named *result* which is the sum of the two function parameters. In the last statement, we have called the function and passed the values for the two parameters. The function will calculate the value of variable *result* by adding the two numbers. We finally get the result shown above.

Note that during our function definition, we specified two parameters, n1 and n2. Try to call the function will either more than two parameters, or 1 parameter and see what happens. Example:

#!/usr/bin/python3

def additionFunction(n1,n2):

    result = n1 + n2

    print('The first number is', n1)

    print('The second number is', n2)

    print("The sum is", result)

additionFunction(5)

In the last statement in our code above, we have passed only one argument to the function, that is, 5. The program gives an error when executed:

```
Traceback (most recent call last):
  File "main.py", line 9, in
    additionFunction(5)
TypeError: additionFunction() missing 1 required positional argument: 'n2'
```

The error message simply tells us one argument is missing. What if we run it with more than two arguments:

#!/usr/bin/python3

def additionFunction(n1,n2):

    result = n1 + n2

    print('The first number is', n1)

    print('The second number is', n2)

    print("The sum is", result)

additionFunction(5,10,9)

We also get an error message:

```
Traceback (most recent call last):
  File "main.py", line 9, in
    additionFunction(5,10,9)
TypeError: additionFunction() takes 2 positional arguments but 3 were given
```

The error message tells us the function expects two arguments but we have passed 3 to it.

In most programming languages, parameters to a function can be passed either by reference or by value. Python supports parameter passing only by reference. This means if what the parameter refers to is changed in the function, the same change will also be reflected in the calling function. Example:

#!/usr/bin/python3

def referenceFunction(ls1):

   print ("List values before change: ", ls1)

   ls1[0]=800

   print ("List values after change: ", ls1)

   return

Calling the function

ls1 = [940,1209,6734]

referenceFunction(ls1)

print ("Values outside function: ", ls1)

The code gives this result:

```
List values before change:  [940, 1209, 6734]
List values after change:   [800, 1209, 6734]
Values outside function:    [800, 1209, 6734]
```

What we have done in this example is that we have maintained the reference of the objects which are being passed and then values have been appended to the same function.

In next example below, we are passing by reference then the same reference will be overwritten inside the same function which has been called:

```
#!/usr/bin/python3

def referenceFunction( ls1 ):

    ls1 = [11,21,31,41]

    print ("Values inside the function: ", ls1)

    return

ls1 = [51,91,81]

referenceFunction( ls1 )

print ("Values outside function: ", ls1)
```

The code gives this result:

```
Values inside the function:  [11, 21, 31, 41]
Vlaues outside function:  [51, 91, 81]
```

Note that the "ls1" parameter will be local to the function "referenceFunction". Even if this is changed within the function, the "ls1" will not be affected in any way. As the output shows above, the function helps us achieve nothing.

Function Parameter Defaults

There are default parameters for functions, which the function creator can use in his or her functions. This means that one has the choice of using the default parameters, or even using the ones they need to use by specifying them. To use the default parameters, the parameters having defaults are expected to be last ones written in function parameters. Example:

#!/usr/bin/python3

```
def myFunction(n1, n2=6):

    pass
```

In above example, the parameter n2 has been given a default value unlike parameter n1. The parameter n2 has been written as the last one in the function parameters. The values for such a function may be accessed as follows:

```
#!/usr/bin/python3

def windowFunction(width,height,font='TNR'):
    # printing everything
    print(width,height,font)

windowFunction(245,278)
```

The code outputs the following:

```
245 278 TNR
```

The parameter *font* had been given a default value, that is, TNR. In the last line of the above code, we have passed only two parameters to the function, that is, the values for width and height parameters. However, after calling the function, it returned the values for the three parameters. This means for a parameter with default, we don't need to specify its value or even mention it when calling the function.

However, it's still possible for you to specify the value for the parameter during function call. You can specify a different value to what had been specified as the default and you will get the new one as value of the parameter. Example:

```
#!/usr/bin/python3
def windowFunction(width,height,font='TNR'):
    # printing everything
    print(width,height,font)
windowFunction(245,278,'GEO')
```

The program outputs this:

```
245 278 GEO
```

Above, the value for parameter was given the default value "TNR". When calling the function in the last line of the code, we specified a different value for this parameter, that is "GEO". The code returned the value as "GEO". The default value was overridden.

Chapter 7- Python Loops

Loops are applicable in situations when we need to perform tasks repetitively. This applies to both when the number of times the task is to be performed and when the number of times is not known. Python supports a number of loops:

For Loop

This loop is used for iterating over something. It will perform something based on each item in the block. The loop is the best if you are aware of the number of times you need the task to be executed.

"RANGE()" function

This function is used when we need to iterate through a sequence of numbers which we specify. The result of the function is an iterator for arithmetic progressions. Open the Python terminal then type the following:

```
>>> list(range(9))
[0, 1, 2, 3, 4, 5, 6, 7, 8]
>>>
```

As shown above, when you list *range(9)*, it will print the values between 0 and 9, with 9 excluded. If the number specified is *n*, then the function usually returns up to *n-1* items, meaning that the list's last item is not included. This can be combined with the *for* loop. Example:

#!/usr/bin/python3

for ax in list(range(9)):

    print (ax)

The code outputs:

```
0
1
2
3
4
5
6
7
8
```

Although 9 is the range specified, it is not included in result.

Note that other than combining *for* loop with *range()* function, it can be used alone. In such a case, you can iterate thought items with the loop. Example, you can iterate through elements of a list with *for* loop:

#!/usr/bin/python3

ls1 = [11,21,31,41]

for ax in ls1:

   print(ax)

We created the list named *ls1* with 4 elements. The *for* loop has been used for iterating through these elements. The code prints the following:

```
11
21
31
41
```

A for loop involves definition of a parameter that will be used for purposes of iteration through elements. In above example, the variable *ax* has been defined and used for iterating through list elements.

The *Range()* function makes the tasks of specifying the range to be executed very easy. You can use the syntax given below:

range(a,b)

The above function will execute and print items between a and b. Practically, consider the example given below:

#!/usr/bin/python3

for ax in range(5, 9):

   print(ax)

The code prints:

5
6
7
8

The code printed values between 5 and 9. Although 5 is included, 9 is not included. This means the initial value is included while the last value is excluded. Also, the range () function takes another parameter that allows us specify the steps by which an increment is to be done. Example:

#!/usr/bin/python3

for ax in range(5, 15, 2):

   print(ax)

The code prints the following:

```
5
7
9
11
13
```

We are printing between 5 and 15, and each iteration will be incremented by 2. Note that 15 is not part of the output.

The *for* loop may also be combined with *else* part. Example:

#!/usr/bin/python3

number = [21,33,53,39,37,75,92,21,12,41,9]

for ax in number:

  if ax%2 == 0:

    print ('There are even numbers in list')

    break

else:

  print ('There are no even numbers in list')

The code will print:

```
There are even numbers in list
```

We used the modulus (%) operator to check whether there are even numbers. The operator returns the remainder after division. If there are numbers in the list in which we remain with 0 after dividing by 2, then the list has some even numbers. Try to create the list without even numbers and see the *else* part will run:

```
#!/usr/bin/python3
number = [21,33,53,39,37,75,93,21,11,41,9]
for ax in number:
   if ax%2 == 0:
      print ('There are even numbers in list')
      break
else:
   print ('There are no even numbers in list')
```

The code will print:

```
There are no even numbers in list
```

While Loop

In *while* loop, we specify a condition to be evaluated after every iteration, and the code will always run provided the condition is true. The execution of code halts immediately the condition becomes false. The loop evaluates the condition after every iteration and the moment it finds itself violating the loop condition, it stops execution of the code. Example:

```
#!/usr/bin/python3
number = 20
```

```
while number < 30:

    print("Value of number is", number)

    number += 1
```

The value of variable *number* was initialized to 20. The *while* condition tests whether this value is below 30. As long as the value of *number* is less than 30, the loop will be executed. The code prints:

```
Value of number is 20
Value of number is 21
Value of number is 22
Value of number is 23
Value of number is 24
Value of number is 25
Value of number is 26
Value of number is 27
Value of number is 28
Value of number is 29
```

As shown, the code counted until the value of *number* was 29. When it reached 30, it found itself violating the loop condition, that is, number must be less than 30. The execution stopped immediately.

Note that 30 is not part of the output. To include it, we can use *less than or equal to* sign (<=) as shown below:

```
#!/usr/bin/python3

number = 20

while number <= 30:

    print("Value of number is", number)

    number += 1
```

The code prints the following:

```
Value of number is 20
Value of number is 21
Value of number is 22
Value of number is 23
Value of number is 24
Value of number is 25
Value of number is 26
Value of number is 27
Value of number is 28
Value of number is 29
Value of number is 30
```

The use of the symbol has included 30 in the output. However, the execution of the program cannot go past that, but it halts immediately it finds itself violating the loop condition. Another example:

#!/usr/bin/python3

age = 15

while (age < 18):

  print ("You are still young, you can't get a personal identity card", age)

  age = age + 1

print ("AFTER THIS YEAR, GO GET A PERSONAL IDENTITY CARD. YOU WILL BE 18 YEARS OLD")

We have specified a default statement to run when the loop condition becomes *false*.

Loop Control

It's possible to change the normal execution of a loop to something else. This can be done using some statements. Once execution has left scope, the objects within that scope will be destroyed. Python supports a number of loop control statements:

Break Statement

This statement helps us terminate execution of a loop prematurely. The execution then begins at the next statement after the loop. It's similar to *break* statement in C. When executing a loop, an external condition may arise that may require instant termination of the loop. The *break* statement can help you in this case. The statement can be used both with *for* and *while* loop. Example:

```
#!/usr/bin/python3

for alphabet in 'Nicholas':    # Example 1
   if alphabet == 'l':
      break
   print ('Current letter is :', alphabet)

number = 5           # Example 2
while number > 0:
   print ('Current variable value :', number)
   number = number -1
   if number == 2:
```

        break

print ("The End!")

The code prints:

```
Current letter is : N
Current letter is : i
Current letter is : c
Current letter is : h
Current letter is : o
Current variable value : 5
Current variable value : 4
Current variable value : 3
The End!
```

First, the loop is iterating through the letters of name *Nicholas*. Once it encounters letter l, it should break or halt iterating through the name letters. In the second example, we are iterating through numbers 5 downwards to 0. When the loop encounters 2, it should break as specified in the condition. Example 2:

In this example, we will be searching through elements of a list. The user is prompted to enter a number which if *found*, the user will get found message. If not, the user will get the *not found* message:

#!/usr/bin/python3

userInput= int(input('Enter number to search: '))

listValue = [11,23,44,39,13,9,8,4,68,21,87]

for ax in listValue:

  if ax == userInput:

    print ('Found')

break

else:

   print ('Not found')

After running the code, search for number 9. You will get this:

```
Enter number to search: 9
Found
```

It's true number 9 is in the list. Search for a number which is not part of the list. Observe the result:

```
Enter number to search: 87763
Not found
```

Continue Statement

With this statement, execution us returned to the start of current loop. Once a loop encounters it, it will begin the next iteration and leave remaining statements in current iteration. It is applicable to both *while* and *for* loops. Example:

#!/usr/bin/python3

for alphabet in 'Nicholas':   # First Example

   if alphabet == 'l':

      continue

   print ('The current Letter is:', alphabet)

number = 5   # Second Example

while number > 0:

number = number -1

if number == 2:

   continue

print ('The current number is :', number)

print ("The End!")

The code prints the following after execution:

```
The current Letter is: N
The current Letter is: i
The current Letter is: c
The current Letter is: h
The current Letter is: o
The current Letter is: a
The current Letter is: s
The current number is : 4
The current number is : 3
The current number is : 1
The current number is : 0
The End!
```

What happened is that the interpreter skipped 1 in the first example and 2 in the second example. This is different from *break* statement.

Pass Statement

This statement is applicable where a statement is needed syntactically but you don't want to execute any statement on that part. It can be seen as *null* operation as nothing happens after it's executed. Example:

#!/usr/bin/python3

for alphabet in 'Nicholas':

```
    if alphabet == 'l':

        pass

    print ('The pass block')

    print ('The current letter is :', alphabet)

print ("The End!")
```

The code gives the following when executed:

```
The current letter is : N
The current letter is : i
The current letter is : c
The current letter is : h
The current letter is : o
The pass block
The current letter is : l
The current letter is : a
The current letter is : s
The End!
```

The code just skipped, but execution resumed to normal after that. You notice that the letter l is now part of the output. This is not what happened in previous two statements.

Chapter 8- Python Classes and Objects

Python is an object oriented programming language. This means that a Python programmer is able to take advantage of the object oriented programming features such as classes.

A class can be defined as a grouping of data and methods which operate on that data. This means that a class has date and methods, whereby, the methods are used for manipulation of the data. The access to the methods of a class is done by use of the dot notation.

Class Definition

To define classes in Python, we use the *class* keyword. This should be followed by a colon. Example:

class testClass():

Once the class has been defined, you can create methods and functions inside it. These will help in data manipulation. Example:

#!/usr/bin/python3

class pythonMaths:

   def add(ax,bx):

     addition = ax + bx

     print(addition)

   def subtract(ax,bx):

     subt = ax - bx

```
    print(subt)

def multiply(ax,bx):

    multiplication = ax * bx

    print(multiplication)

def division(ax,bx):

    div = ax / bx

    print(div)
```

We have defined a class named *pythonMaths*. The class a number of methods. To access any of these methods, you must use class name, the dot (.) and the method name. Example:

To access the add method in above *pythonMaths* class, type the following on Python terminal:

pythonMaths.add(2,3)

Note that the class name comes first, followed by the method name then the parameters inside parenthesis. The function expects two parameter values, that is, values for parameters ax and bx. If you pass values for more than two parameters, or even one parameter, then an error will be returned.

Note that everything in the class has been indented. This should always be the case. If you don't, an error message will be generated.

The class methods may also be called from within the class itself. This calls for us to create an instance of the class which will be used for accessing the class methods. Example:

#!/usr/bin/python3

```python
class class2():
    def firstMeth(self):
        print("The first method")
    def secondMeth(self,aString):
        print("Second method, string alongside:" + aString)
def main():
    # instantiate class and call methods
    c = class2 ()
    c.firstMeth()
    c.secondMeth(" We are now testing")
if __name__ == "__main__":
    main()
```

The code prints the following when executed:

```
The first method
Second method, string alongside: We are now testing
```

The argument *self* is normally to refer to object itself. That's why we use the word, and it's a keyword in Python. When used inside a method, *self* refers to a specific instance of the object being operated on. Whenever you see the keyword *self* in Python, know it refers to first parameter of the instance methods. It is used for accessing member objects. However, you notice that when calling the two methods in our code, that is, *firstMeth()* and *secondMeth()*, we never specified the self keyword as Python does this for us. After calling an instance method, Python knows how to automatically pass the *self* argument whether it has been provided or not. This means you may choose to provide it or not. We created an instance of the class *class2* and the instance was named *c*. This was done in the following line:

c = class2 ()

The c is an object of class *class2*. This means we can use the object to access all methods and properties of the class. You only have to care about the non-self arguments. Notice how a string was appended to initial text in *secondMeth*.

Built-in Attributes

There are some built-in attributes which are kept by all classes and to access them, we use the dot operator similar to the other attributes. These include the following:

__dict__: This is a dictionary with the namespace for the class.

__doc__: The class documentation string or none, in case it is not defined.

__name__: The name of the class.

__module__: The name of the module in which the class has been defined. In the interactive mode, the attribute becomes "__main__".

__bases__: This is a tuple, possibly empty, having base classes, added in the order that they occur in your base class list.

Example:

#!/usr/bin/python3

class Worker:

'The base class. Its common to all instances'

workerCount = 0

def __init__(self, name, age):

  self.name = name

  self.age = age

  Worker.workerCount += 1

def showCount(self):

  print ("The total number of workers is %d" % Worker.workerCount)

def showWorker(self):

  print ("Name : ", self.name, ", Age: ", self.age)

worker1 = Worker("Gishon", 26)

worker2 = Worker("Esther", 24)

print ("Worker.__doc__:", Worker.__doc__)

print ("Worker.__name__:", Worker.__name__)

print ("Worker.__module__:", Worker.__module__)

print ("Worker.__bases__:", Worker.__bases__)

print ("Worker.__dict__:", Worker.__dict__)

The code prints the following when executed:

```
Worker.__doc__: The base class. Its common to all instances
Worker.__name__: Worker
Worker.__module__: __main__
Worker.__bases__: (,)
Worker.__dict__: {'__module__': '__main__', '__doc__': 'The base class. Its common to all instances', 'workerCount': 2, '__init__
```

Garbage Collection

Sometimes, the memory may be occupied by objects that are no longer needed. Python clears them from the memory automatically, a process known as *garbage collection*. This way, Python is able to reclaim blocks of memory that are no longer in use. The garbage collector is launched when a program is executed and it runs once a reference count to an object has reached a zero. The reference count to an object changes with change in the number of aliases pointing to it.

The reference count to an object increases when a new name is assigned or when it's added into a container such as tuple, list or dictionary. Once the del statement is used to delete the object, the value of count will decrease, or once its reference has gone out of scope or once the reference is reassigned.

Example:

ax = 5   # object created

bx = ax   # Increase the ref. count for <5>

bz = [bx]   # Increase the ref. count for <5>

del ax   # Decrease the ref. count for <5>

bx = 70   # Decrease the ref. count for <5>

bz[0] = -1 # Decrease the ref. count for <5>

One is not capable of noticing once the garbage collector has destroyed an orphaned instance. However, in Python, it is possible for a class to implement a destructor named "__del__()" which will be invoked when a particular object is almost destroyed. Any non memory resources which are not being used by an instance can be cleaned by use of this method. The __del__() destructor normally shows the class name for the instance that is almost being destroyed.

Example:

#!/usr/bin/python3

class Region:

  def __init__(self, ax=0, bx=0):

    self.ax = ax

    self.bx = bx

  def __del__(self):

    class_name = self.__class__.__name__

    print (class_name, "already destroyed")

rg1 = Region()

rg2 = rg1

rg3 = rg1

print (id(rg1), id(rg2), id(rg3));   # to print object IDs.

del rg1

del rg2

del rg3

The code prints:

```
139921453750144 139921453750144 139921453750144
Region already destroyed
```

The best idea for you is to create your classes in some separate files. You can then use the "import" statement so as to import these classes into your main program. Suppose the code given above was created in the file "Region.py" and it has no executable code, then we can do this as follows:

#!/usr/bin/python3

import region

rg1=region.Region()

Inheritance

In Python, you don't have to create your class from scratch but you can inherit from ma certain class, normally known as the "parent" class. The parent class should be place in parenthesis after the definition of the new class.

Since the parent class has some attributes, the new class, which is the child class will be allowed to use these attributes in such a manner that they have been defined in the child class. It is also possible for the child class to override the methods and the data members from the parent class.

Python inheritance takes the following syntax:

class DerivedClassName(BaseClassName):

  derived_class_body

Example:

```python
#!/usr/bin/python3
# Example file for working with classes
class parentClass():
  def firstMeth(self):
    print("The first method in parentClass")
  def secondMeth(self,aString):
    print("We are testing" + aString)
class childClass(parentClass):
  #def firstMeth(self):
    #parentClass.firstMeth(self);
    #print "firstMeth for Child Class"
  def secondMeth(self):
    print("childClass secondMeth")
def main():
  # exercising class methods
  c = childClass()
  c.firstMeth()
  c.secondMeth()
if __name__ == "__main__":
```

main()

The code prints the following when executed:

```
The first method in parentClass
childClass secondMeth
```

In the *childClass*, we have not defined the *firstMethod* but we have obtained it from parent class. That is how inheritance works in Python. The child class has inherited a method from the parent class.

Another Example:

```
#!/usr/bin/python3

class Worker:
    'A common class to all the workers'
    workerCount = 0

    def __init__(self, name, wage):
        self.name = name
        self.wage = wage
        Worker.workerCount += 1

    def showCount(self):
        print ("Total Workers %d" % Workers.workerCount)

    def showWorker(self):
        print ("Name : ", self.name,  ", Wage: ", self.wage)

#Creating first object of Worker class"
```

worker1 = Worker("Bosco", 2500)

#Creating second object of Worker class"

worker2 = Worker("June", 3000)

worker1.showWorker()

worker2.showWorker()

print ("Total Workers %d" % Worker.workerCount)

The above code clearly demonstrates how you can create an instance of a class and use it to access members or methods of the parent class. It gives the following result once executed:

```
Name :  Bosco , Wage:  2500
Name :  June  , Wage:  3000
Total Workers 2
```

We have created two instances of the class Worker, that is, worker1 ad worker2. Each of these instances is a worker, the first one Bosco and the second one June. We have used these instances to access the showWorker method defined in the class. This method returns the name and the wage for the worker.

Multiple Inheritance

In Python, one can inherit from more than one class at once. This is not the case with other languages like Java and C#. Python's multiple inheritance takes this syntax:

Class Childclass(ParentClass1, ParentClass2, ...):

  # the initializer

the methods

Example of Python multiple inheritance:

```python
#!/usr/bin/python3
class ParentClass1():
 def superMethod1(self):
  print("Calling superMethod1")
class ParentClass2():
 def superMethod2(self):
  print("Calling superMethod2")
class ChildClass(ParentClass1, ParentClass2):
 def childMethod(self):
  print("The child method")
ch = ChildClass()
ch.superMethod1()
ch.superMethod2()
```

The code will print:

```
Calling superMethod1
Calling superMethod2
```

We defined two methods, one in first Super Class ad the second one in second Super Class. The child class has then inherited from these two classes. It has accessed the methods that have been defined in these two classes. That is how we can inherit from more than one classes in Python.

Python Constructors

A constructor refers to a class function for instantiating an object to some predefined values. It should begin with a double underscore (__). It is the __init__() method. Example:

```
#!/usr/bin/python3
class Worker:
    workerName = ""
    def __init__(self, workerName):
        self.workerName = workerName
    def sayHello(self):
        print("Welcome to our company, " + self.workerName)
Worker1 = Worker("June")
Worker1.sayHello()
```

The code prints the following when executed:

```
Welcome to our company, June
```

What we have done is that we have used a constructor to get the name of the user.

Overriding Class Methods

When coding in Python, we are allowed to override methods that are defined in parent class. You may need to have a different functionality in the child class, and this is a good reason for overriding a parent method. To override the method, we only have to pass different arguments to it as demonstrated below:

```python
#!/usr/bin/python3

class ParentClass:      # define the parent class
    def firstMethod(self):
        print ('A call to parent method')

class ChildClass(ParentClass): # define the child class
    def firstMethod(self):
        print ('A call to child method')

c = ChildClass()        # An instance of the child class
c.firstMethod()         # The child calls the overridden method
```

The code will print the following once executed:

```
A call to child method
```

The method named *firstMethod* had been defined in the parent class. The function has been redefined in the child class but this time, it prints a different text than what it was printing in parent class. We have achieved method overriding.

Example 2:

#!/usr/bin/python3

```
class AB():
    def __init__(self):
        self.__ax = 2

    def method1(self):
        print("method1 from class AB")

class BC(AB):
    def __init__(self):
        self.__bx = 2

    def method1(self):
        print("method1 from class BC")

bc = BC()
bc.method1()
```

The code prints the following once executed:

```
method1 from class BC
```

Operator Overloading

As you know, the + operator can be used for addition of numbers as well as for concatenation of strings. The reason is that the operator has been overloaded by both the *str* and *int* classes. The operators are methods that have been defined in their respective classes. Definition of methods for the operators is referred to as *operator overloading*. For the + operator to be used with custom objects, the method __add__ should be defined. Example:

```
#!/usr/bin/python3
import math
class CircleClass:
  def __init__(self, circleRadius):
   self.__circleRadius = circleRadius
  def setRadius(self, circleRadius):
   self.__circleRadius = circleRadius
  def getCircleRadius(self):
   return self.__circleRadius
  def area(self):
   return math.pi * self.__circleRadius ** 2
  def __add__(self, another_circle):
   return CircleClass( self.__circleRadius + another_circle.__circleRadius )
circle1 = CircleClass(2)
print(circle1.getCircleRadius())
circle2 = CircleClass(3)
print(circle2.getCircleRadius())
```

circle3 = circle1 + circle2 # The + operator has been overloaded by adding the method __add__

print(circle3.getCircleRadius())

The code prints the following once executed:

2
3
5

What we have done is that we have added the __add__ method that has helped us add some two circle objects. Inside the method, a new object was created then returned to the caller. There are numerous special methods just like the __add__. They include:

- \+   __add__(self,other)   For Addition
- \-   __sub__(self,other)   Subtraction
- *   __mul__(self,other)   Multiplication
- %   __mod__(self,other)   Returns the Remainder
- <   __lt__(self,other)   For Less than
- /   __truediv__(self,other) For Division
- <=   __le__(self,other)   For Less than/equal to
- !=   __ne__(self,other)   For Not equal to
- ==   __eq__(self,other)   For Equal to
- >   __gt__(self,other)   For Greater than
- [index] __getitem__(self,index)   The Index operator
- >=   __ge__(self,other)   For Greater than/equal to
- in   __contains__(self,value)   Checks the membership
- str   __str__(self)   For string representation
- len   __len__(self)   Checks number of the elements

The code given below makes use of above functions for operator overloading:

```python
#!/usr/bin/python3
import math
class CircleClass:
    def __init__(self, circleRadius):
        self.__circleRadius = circleRadius
    def setRadius(self, circleRadius):
        self.__circleRadius = circleRadius
    def getRadius(self):
        return self.__circleRadius
    def circleArea(self):
        return math.pi * self.__circleRadius ** 2
    def __add__(self, second_circle):
        return CircleClass( self.__circleRadius + second_circle.__circleRadius )
    def __gt__(self, second_circle):
        return self.__circleRadius > second_circle.__circleRadius
    def __lt__(self, second_circle):
        return self.__circleRadius < second_circle.__circleRadius
    def __str__(self):
```

```
    return "Circle has a radius of " + str(self.__circleRadius)
```

```
circle1 = CircleClass(2)

print(circle1.getRadius())

circle2 = CircleClass(4)

print(circle2.getRadius())

circle3 = circle1 + circle2

print(circle3.getRadius())

print( circle3 > circle2) # We have added __gt__ method, hence this is possible

print( circle1 < circle2) # we added __lt__ method, hence this is possible

print(circle3) #we added __str__ method, hence this is possible
```

The code prints the following:

```
2
4
6
True
True
Circle has a radius of 6
```

Chapter 9- Exception Handling

With Exception handling, we are able to detect errors and handle them appropriately. If you are searching for a file and it is not found for example, you can raise an error message. The *try* and *except* statements are used in Python for error handling. These statements follow the same concept followed in the *if-else* statement, in which if the *if* part runs, the *else* part will not run. Consequently, if the *try* part runs, the *except* part won't run. If the *try* part fails, then the exception part will run with error generated in *try* part. With exception handling, your code can be kept running even in cases when it could have failed. Error handling is also a good way of logging any problems that you may have in your code. You may also correct the problem with your code.

The *try* and *except* takes this syntax:

try:

   # Add code

   # to throw an exception

except <ExceptionType>:

   # The exception handler to alert user

To see it work, you only have to write the code that will throw an exception. In case of occurrence of an exception, the *try* code will be skipped, If you have a matching exception in *except* part, then it will be executed to handle the exception.

Example:

#!/usr/bin/python3

```
try:

    fl = open('filename.txt', 'r')

    print(fl.read())

    fl.close()

except IOError:

    print('The file was not found')
```

The code will print:

```
The file was not found
```

We are trying to access a file and read it. That is in the *try* statement. However, in the *except* part, we have the *IOError*, which handles input/output exceptions. We have defined what should happen in case of such an occurrence, that is, if the file is not found. It should execute the *print* statement. Just run the code and ensure you don't have the file. The *print* statement will be printed.

If the file is found, then the part under *except* statement will be skipped. In my case, the exception occurred, hence the *try* part was skipped, that is, the file was not read. For the *except* part to run, the exception that occurs must match the one you are handling. Note that our code given above is only capable of handling the "IOError" exception. To handle ay more errors, we should add other *except* clauses. This means that we may have numerous except clauses in a single *try* clause, as well as an optional *else* or finally clause.

The following syntax should be followed:

try:

 <try body>

except <Exception1>:

 <Exception handler1>

except <ExceptionN>:

 <Exception handlerN>

except:

 <Exception handler>

else:

 <else body>

finally:

 <finally body>

The *except* works similarly to *elif* clause. After the occurrence of an exception, it is checked to know that except that matches. If a match is found, then it's executed. Note that in our last *except*, we don't have *ExceptionType*. This means if the exception doesn't match any of the exception types, Note that the statements below the *else* part will only run after none of the exceptions is raised. The statements in the *finally* clause will always run whether an exception is raised or not.

Example:

#!/usr/bin/python3

```
try:

    n1, n2 = eval(input("Type numbers and separate then with a comma : "))

    answer = n1 / n2

    print("The result is", answer)

except ZeroDivisionError:

    print("Division by zero gives error !!")

except SyntaxError:

    print("Comma not found. Type numbers and separate then with a comma as 1, 2")

except:

    print(" A wrong input was found")

else:

    print("Exceptions not found ! ")

finally:

    print("This part for finally will always run")
```

You can enter the two numbers as instructed then fail to separate them with a comma. An exception will be raised. Whether you separate then with a comma or not, you notice the *finally* clause will always run.

With the *eval()* function, a Python program is capable of running the Python code in itself. The function should be supplied with string argument.

Raising Exceptions

If you are in need of raising exceptions from your own methods, you should use the *raise* keyword. Example:

raise ExceptionClass("An argument")

Example:

```
#!/usr/bin/python3
def getAge(yourAge):
 if yourAge < 0:
  raise ValueError("Your age MUST be a positive integer value")
 if yourAge % 2 == 0:
  print("Your age is an even number")
 else:
  print("Your age is an odd number")
try:
 number = int(input("What's your age? "))
 getAge(number)
except ValueError:
 print("Only positive integers are allowed")
except:
 print("something went wrong")
```

Run the code. You will be prompted to enter your age. If it's an even number, you get the following:

```
What's your age? 26
Your age is an even number
```

This is because after dividing 26 by 2, the remainder is 0. This means it's an even number. Enter an odd number for the age. You will get the following:

```
What's your age? 25
Your age is an odd number
```

Enter a negative integer for the age ad see what happens. You will get the following:

```
What's your age? -20
Only positive integers are allowed
```

That is how you can raise exceptions on your methods.

Exception Objects

Now that you are familiar with handling exceptions in Python, let us learn how to access the exception object in the code for exception handler. The following code may be used for assigning a variable to an exception object. The following syntax should be used for this:

try:

   # the code to throw an exception

except TypeOfException as ex:

   # code for handling the exception

The exception object can be stored in the variable ex. The exception can then be used in the exception handling code. Example:

#!/usr/bin/python3

try:

n1 = eval(input("Type a number: "))

print("You entered ", n1)

except NameError as ex:

print("Exception:", ex)

Run the code then type a number when prompted to do so. You will get the following:

```
Type a number: 8
You entered  8
```

As shown above, the code executed correctly. Ow, run the code then enter a string:

```
Type a number: number
Exception: name 'number' is not defined
```

As shown above, entering a string raises an exception. This is because the code expected you to enter a number but you have entered a string. This raises an error.

Custom Exception Class

It is possible for you to create some custom exception class. This requires you to extend the BaseException class or a subclass of the BaseException class. The BaseException class can be seen as the root of all the exception classes in Python.

In your Python text editor, create a new file named *NegativeAgeException.py* then add the following code to it:

```
class NegativeAgeException(RuntimeError):
    def __init__(self, yourAge):
        super().__init__()
        self.yourAge = yourAge
```

What the code does is that it creates a new exception class named NegativeAgeException. The class has only one constructor that will call the parent class constructor by use of *super().__init__()* then set the value of *yourAge* argument. The custom exception class can be used as follows:

```
#!/usr/bin/python3
def getAge(yourAge):
    if yourAge < 0:
        raise NegativeAgeException("Age MUST be a positive integer")
    if yourAge % 2 == 0:
        print("The age is an even number")
    else:
        print("The age is an odd number")
try:
    n = int(input("What's your age? "))
```

getAge(n)

except NegativeAgeException:

print("Enter a positive integer")

except:

print("something went wrong")

You can run the code and enter a numeric value, a positive one. You will get this:

```
What's your age? 78
The age is an even number
```

The code runs correctly. Now run the code then enter a negative integer as the value for your age:

```
What's your age? -23
Traceback (most recent call last):
  File "C:/Users/admin/age.py", line 14, in <module>
    getAge(n)
  File "C:/Users/admin/age.py", line 5, in getAge
    raise NegativeAgeException("Age MUST be a positive integer")
NameError: name 'NegativeAgeException' is not defined

During handling of the above exception, another exception occurred:

Traceback (most recent call last):
  File "C:/Users/admin/age.py", line 15, in <module>
    except NegativeAgeException:
NameError: name 'NegativeAgeException' is not defined
>>>
```

The exception will be raised as shown above.

Let us create an exception related to *RuntimeError*. We will creat6e a class to be a subclass of *RuntimeError* class. It is a good way of getting more information after the occurrence of an exception. The exception is raised in the try block then handled in the except block. We will use the variable ex for creation of an instance of class *Networkerror*.

```
class Networkerror(RuntimeError):
    def __init__(self, argu):
        self.args = argu
```

After defining the class as above, the exception can be raised as follows:

```
try:
    raise Networkerror("Wrong hostname")
except Networkerror,ex:
    print ex.args
```

Chapter 10- Python Modules

The purpose of modules is to help us organize our code in a logical way. When sections of related code are grouped, it becomes easy to understand and use the code. A module comes with a number of attributes that one can bind as well as reference. The module is simply a file that has Python code. It can be used for definition of classes, functions and variables. The module may also have code that can run.

Example:

def print_func(par):

  print "Hello : ", par

  return

If you need to access and use code for another Python source file, you can use the *import* keyword to have it in the source file you are working on. That is how modules are used in other source files:

import module1[,module2[,... moduleN]

Once the Python interpreter encounters the *import* statement, it will search for the specified module in the path and import it into your current source file. The search path is made up of a number of directories that the interpreter must search whenever it needs to import a module.

Python comes with the default *hello.py* file with the basic *Hello* code. For us to use this, we must import the support module by adding the *import* statement at the top of our script. Example:

#!/usr/bin/python3

\# Importing a module named support

import support

\# Let us call a method defined in the module

support.print_func("Nicholas")

What we have done is that we have used the *import* keyword to import the module named support. The module is simply written in a file named *support.py*. The module has a function named *print_func*. We have then called this method in our above file. Note the syntax used for calling methods defined in other modules. We begin by the module name then the method name, joined/separated using a dot (.)

Note that a module will only be imported once regardless of the number of times you call it via the *import* keyword. This helps in prevention of execution of a module repeatedly.

Sometimes, you may not be in need of importing or using the entire module. Some modules are also heavy and may occupy too much memory space. The *from* keyword helps you import only a number of attributes from the Python modules. This means that the entire module won't be imported but only the attributes you have specified. Syntax:

from moduleName import attribute1[, attribute2[, ... attributeN]]

Example:

Python has a module named *fib*. We can import the *Fibonacci* method from this module as follows:

#!/usr/bin/python3

\# Python module for Fibonacci numbers

```
def fib(num): # return then Fibonacci series to num
    answer = []
    ax, bx = 0, 1
    while bx < num:
        answer.append(bx)
        ax, bx = bx, ax + bx
    return answer
```

Now that we have the module above, that is, fib, we can import its attribute as follows from the Python interactive interpreter.

>>>from fib import fib

>>>fib(50)

Note that you must save the code with the name *fib.py* to designate it as a Python file. The file should also be saved in a directory known to the Python interpreter. This will make it easy for the interpreter to search for the file, open it and get the attribute or method you need to import. In my case, I have saved the file as *fib.py*. Once I compile it, I get no error. It is after that I open the Python interactive terminal then I import the attribute and call the method as shown below:

```
>>> from fib import fib
>>> fib(50)
[1, 1, 2, 3, 5, 8, 13, 21, 34]
>>>
```

The first statement helps us import the *fib* function/method from the module. We are then able to call the method and pass it an argument to it.

It successful returns the fibonacci of 50 to us. Note that we didn't import the entire module, but only an attribute from it.

Also, it is possible for one to import all attributes of a module into current namespace by use of this statement:

from moduleName import *

The above provides us with an easy way of importing all named contained in a module into current workspace.

Inside the module, the name for the module is provided in string form and as value of global variable __name__. The module code will be run in similar way as you had imported it, but the __name__ will be set to __main__.

You can modify the code for your *Fibonacci* module to appear as follows. You only add a section of code at its end:

#!/usr/bin/python3

Python module for Fibonacci numbers

def fib(num): # return then Fibonacci series to num

    answer = []

    ax, bx = 0, 1

    while bx < num:

      answer.append(bx)

      ax, bx = bx, ax + bx

    return answer

```
if __name__ == "__main__":

    fb = fib(50)

    print(fb)
```

You can run the code and see what it prints. It will give you the following result:

```
[1, 1, 2, 3, 5, 8, 13, 21, 34]
```

As shown above, the result is simply the Fibonacci of 50. This time, we did not have to call the module from the Python terminal but we have done directly in the module code.

Locating Modules

Once you attempt to import a module via the *import* statement, there are a number of directories that the Python interpreter searches to locate it. First, the Interpreter must search for the module in the current directory. If it is not found, it proceeds to search in all the directories in the shell variable, that is, PYTHONPATH. If the interpreter doesn't find the module, it proceeds to check for it in the default path. In UNIX systems, the default path is normally located at /usr/local/lib/python3/.

The search path for the module is kept in system module sys in the form of the variable *sys.path*. The variable, that is, *sys.path*, has the current directory, the PYTHONPATH and python-dependent default.

PYTHONPATH is simply an environment variable. It is made up of numerous directories. It has same syntax as that of shell variable PATH. Example of PYTHONPATH in Windows:

set PYTHONPATH =c:\python34\lib;

Example of PYTHONPATH in UNIX:

set PYTHONPATH =/usr/local/lib/python

Namespaces and Scope

Variables are simply names mapping to objects. The *namespace* refers to dictionary of variable names (the keys) together with the corresponding objects (the values). Python statements are allowed to access both global and local variables. This brings the concept of *local namespace* and *global namespace*. In case same name is used for both a local and global variable, the global variable will be shadowed by the local variable.

Every function has a local namespace. Class methods normally follow similar scoping rule just like ordinary functions. Python is capable of making wise guesses regarding whether a variable is local or global. Any variable assigned a value is assumed to be local. If you need to assign a value to some global variable within a function, you should use *global* keyword. Example:

global VariableName

The above tells Python interpreter that VariableName is a global variable. The Python interpreter will not search for the local namespace of the variable. Example:

#!/usr/bin/python3

Wage = 3000

def IncreaseWage():

Uncomment the below line and fix your code:

#global Wage

Wage = Wage + 200

print ("The initial value for wage is:", Wage)

IncreaseWage()

print ("The value of wage after increase is:", Wage)

The above code generates an error message. Since *Wage* was assigned a value, it was assumed that it's a local variable. However, it was defined in the global namespace. We however access value of local variable Wage without setting its value. An UnboundLocalError was generated. To fix the problem, uncomment your global statement to remain with this:

#!/usr/bin/python3

Wage = 3000

def IncreaseWage():

Uncomment the below line and fix your code:

global Wage

Wage = Wage + 200

print ("The initial value for wage is:", Wage)

IncreaseWage()

print ("The value of wage after increase is:", Wage)

The error will be removed, and the code will print the following once executed:

```
The initial value for wage is: 3000
The value of wage after increase is: 3200
```

dir() Function

This is an in-built function that returns strings that have been sorted with names a module has defined. The list shows all names, functions and variables the module has defined. Example:

#!/usr/bin/python3

Import module math which is built-in

import math

cont = dir(math)

print (cont)

The code will print the following:

```
['__doc__', '__file__', '__loader__', '__name__', '__package__', '__spec__', 'acos', 'acosh', 'asin', 'asinh', 'atan',
```

The output is just a section of the output as the code prints a list of functions, variables and names defined by the *math* module. See how the *dir()* function was called with the argument to the function being the name of the module.

locals () and global() Functions

These two function are useful for returning names in local and global namespaces based on location you have called them from. If you call *locals()* from within a function, it returns a list of names accessible from within that function. If you call *globals()* from within a function, it returns a list of names accessible globally from within the function.

Note that both have a dictionary as the return type. This means to extract the names, we can use the *keys()* function.

reload () Function

This is another function in Python. After you have imported a module into a script, the module's code in top-level portion will be executed for only once. However, there might be situations in which you will wat to run this top-level code for more than once. In such a case, you call the *reload()* function. The function works by importing a module that had been imported. The *reload()* function has the following syntax:

reload(module_name)

The *module_name* denotes the name of the module that you need to reload. However, it's not the string that has the name of the module. If you need to reload a module named *hello* for example, use this:

reload(hello)

A package refers to a file director organized in a hierarchical structure. It represents a single application with modules, sub-packages and even sub-subpackages.

Chapter 11- File Handling

With Python, you can access your files in the system and read them, write to them and even modify their contents. You are now aware on how to read and write to standard output. I will be showing you how to do this to your files.

There are default Python functions that can be used for file handling. The *file* object can help you do much of the calculation. The *file* object can help you do the manipulation on your files.

open() Function

For a file to be read, written to or even modified, it must first be opened. This is done using the Python in-built function named *open()*. When invoked, the function will create *file* objects that can be used for calling support methods associated with it. Here is the method's syntax:

file objectName = open(file_name [,access_mode][,buffering])

The *file_name* is a string representing the name of the file to be opened. For the *access_mode*, the file can be opened for read, write or append. The default mode foe opening the file into is *read (r)*. If the value for buffering is set to 0, then no buffering will be done. If it is set to 1, line buffering will be done after accessing the file. If you specify another integer greater than 1 for buffering, then buffering will be done at the size that you have specified. The integer is normally taken as buffer size. If it is set to a negative integer, the default buffering size for the system is used.

There are different modes in which the file may be opened. They include:

- r- the file is opened for reading only, and it's the default mode. The file pointer is placed at the start of the file.
- rb- the file is opened I binary format for reading only. The file pointer is placed at the start of the file.
- r+- the file is opened for both reading and writing. The file pointer is placed at the start of the file.
- rb+- the file is opened in binary format for reading and writing. The file pointer is placed at the start of the file.
- w- the file is opened for writing only. If the file exists, it is overwritten. If the file does not exist, a new one is created.
- wb- the file is opened in binary format for writing only. If the file exists, it is overwritten. If the file doesn't exist, a new one is created for both reading and writing.
- wb+- the file is opened in binary format for both reading and writing. If the file exists, it is overwritten. If the file doesn't exist, a new one is created.
- a- the file is opened for appending. If the file exists, the file pointer is placed at the end of the file. If the file doesn't exist, a new one is created for writing.
- ab- the file is opened in binary format for appending.
- a+- the file is opened for both reading and appending. If the file exists, the pointer is moved to the end of the file. This puts the file in the append mode. If the file doesn't exist, a new one is created for writing and reading.
- ab+- the file is opened in binary format for both reading and appending. If the file exists, the pointer is placed at the end of the file. The file is kept in append mode. If the file doesn't exist, a new one is created for writing and reading.

The file object is related to these attributes:

- file.closed- it will return *true* if file is closed, and *false* otherwise.
- file.mode- this returns the mode in which the file is opened.
- file.name- this will return the file name.

Example:

#!/usr/bin/python3

Opening the file

f = open("names.txt", "wb")

print ("The file name is: ", f.name)

print ("If the file closed? ", f.closed)

print ("Which mode is the file in? ", f.mode)

close the file

f.close()

Ensure that you have the file named *names.txt* in the directory then run the code. The code will give result based on the file. In my case, it prints the following:

```
The file name is:  names.txt
If the file closed?  False
Which mode is the file in?  wb
```

We have the name of the file which was obtained by calling the *name* attribute. The *false* in the result tells us that the file is not closed. Also, it is clear that the file has been opened in binary format for writing.

close() Method

This method should be called for closing a file. It first flushes the unwritten information then closes the file. Once the file has been closed, no further writing can be done. If reference object for file is assigned to some other object, Python will automatically close the file. Whenever you need to close a file, call the *close()* method. The method syntax is as follows:

fileObject.close();

Example:

#!/usr/bin/python3

Opening the file

f = open("names.txt", "wb")

print ("The file name is :", f.name)

Close the opened file

f.close()

print("Is the file closed ? ", f.closed)

Run the code in the directory with the file names.txt and the following result will be printed:

```
The file name is : names.txt
Is the file closed ?  True
```

In our previous example, we got *FALSE* when we called the *f.closed* property. This meant that the file was not closed. The reason is that we had not called the *close()* method. In above example, we have called close() method on our file object. This closed the file, hence we get True after calling the f.closed property. This means that file has been closed.

write() Method

This Python method helps us write to files. For you to use it, first open the file then pass the string to be written to the file. Note that Python strings may have binary data rather than strings only. Note that the method doesn't add newline character (\n) at the end of your string. The method has the following syntax:

fileObject.write(string);

The parameter to the method is the content to be written to the file. Once the writing has been completed, call the *close()* method to close the file.

Example:

#!/usr/bin/python3

Opening the file in write mode

f = open("names.txt", "w")

f.write("This is the first line we are writing into the file.\nThis is the second line we are writing into the file!!\n")

Close the opened file

f.close()

Run the code in the directory with the file *names.txt*. You will notice that the specified strings will be written/added to the file. We first opened the file in *write* mode. We then called the write() method and passed to it the two strings to be written into the file. Each string has been enclosed into its quotes. To start a new line after writing the first string, we have used the newline character, that is, \n. The following will be written to the file *names.txt* after running the above code:

This is the first line we are writing into the file.

This is the second line we are writing into the file!!

read () Method

This method helps us read a string from an open file. Note that the file may have either textual data or binary data. Both can be read via this function. The method syntax is as follows:

fileObject.read([count]);

The parameter to the function is the number of bytes that you need to read from the file. The method usually begins to read from the beginning of the file. If you don't specify a value for *count*, then the method will read from the file as much as it can. Most probably, the method will read till the file's end.

Example:

#!/usr/bin/python3

Opening the file

f = open("names.txt", "r+")

txt = f.read(10)

print ("The method read the string : ", txt)

Closing the opened file

f.close()

In my case, the following text was read from the file:

```
The method read the string :    This is th
```

Note that we had instructed the method to read only 10 bytes from the file, and that is why not all the file contents were read.

File Positions

You may use *tell()* method to tell current position in a file. This tells where the next read or write will start from the next time you attempt to do so on the file.

If you need to change this position, you can use the *seek(offset[, from])* method. The argument, that is, offset specifies the number of bytes that should be moved. The argument *from* specifies reference position from which bytes should be moved.

If the value of *from* is 0, then the reference position is the starting point of the file. If you set it to 1, then the current position will be used as reference position. If it is set to 2, the file's end will be used as reference position.

Example:

#!/usr/bin/python3

Openig the file

f = open("names.txt", "r+")

str = f.read(10)

print ("The function read the string : ", str)

Checking the current position

pos = f.tell()

print ("The current position for the file is : ", pos)

Reposition the pointer to the beginning

pos = f.seek(0, 0)

str = f.read(10)

print ("The read String again is : ", str)

Close the opened file

f.close()

Run the code from the directory you have stored the file names.txt. In my case, it returns the following:

```
The function read the string :   This is th
The current position for the file is :   10
The read String again is :   This is th
```

You notice that in both case, the same string was read. We first read the first 10 bytes of the file. This moved the position in the file to 10 as shown in the output. We then called the *seek()* function to reset the position of the file to the beginning. When we issue the read command, it again reads from the beginning, hence we get the same output.

rename() Method

This method helps in renaming file names, and it takes two arguments. The method takes two arguments, the first one being the current name of the file and the second one being the new name to be given to the file. Note that this method is provided by a Python module named *os*. For you to use the function, you must first import the module. The method takes the following syntax:

os.rename(current_filename, new_filename)

Example:

#!/usr/bin/python3

import os

Rename the file from names.txt to mytext.txt

os.rename("names.txt", "mytext.txt")

We began by importing the *os* module via the *import* keyword. It is after that we have called the *rename()* method. Note the syntax used for calling the method. We began by the module name, that is, os, then the method name, that is, rename(). Two arguments were passed to the method. The first one is the name of the file we need to rename which is *names.txt*. We have then defined the new name we need to give to the file, that is, *mytext.txt*. That is how files should be renamed in Python.

remove() Method

This method can be used for deletion of a file. The method is called and the name of the file to be deleted or removed is passed as the argument. Again, this method is provided in the *os* module, hence you must first import the module before using the method. The syntax for the method is as follows:

os.remove(file_name)

Example:

#!/usr/bin/python3

import os

Deleting the file named mytext.txt

os.remove("mytext.txt")

In the above case, we first imported the *os* module into the script. We have then called the *remove()* method from this module. Again, we used the same syntax to call the method as we did in our previous example. The name of the file to be deleted is *mytext.txt*, hence this has been passed as the argument to the function.

mkdir() Method

Files are kept in directories. The *os* module comes with several methods that can be used for working with directories. The *mkdir()* command provided in the *os* module helps in creating directories in your current directory. The method expects an argument to be passed to it, and this should be the name of directory to be created. Its syntax involves calling the *os* module first as shown below:

os.mkdir("newdirname")

Example:

#!/usr/bin/python3

import os

Create the directory "testdirectory"

os.mkdir("testdirectory")

In the above example, we have created a directory named *testdirectory*. The name of the directory has been passed as the argument to the method.

chdir() Method

This method helps in changing the current directory. It takes one argument, which is the name for the directory you need to shift or change to. The syntax for the method is as follows:

os.chdir("newdirname")

Example:

#!/usr/bin/python3

import os

Change directory to "/home/directory1"

os.chdir("/home/directory1")

In above example, we have simply changed directory to *directory1*.

getcwd() Method

This method returns the current working directory. The method is defined in os module, hence it takes the following syntax:

os.getcwd()

Example:

```
#!/usr/bin/python3

import os

# This will return location of current directory

print(os.getcwd())
```

The code will return the current working directory for the user. Notice that the *os.getcwd()* method has been called within the *print()* method. This will help in displaying the current working directory.

rmdir() Method

This method helps in deleting a directory that is passed to it in the form of an argument. Before a directory can be removed, all its contents should first be deleted. Here is the syntax for the method:

os.rmdir('dirname')

When deleting a directory, a fully qualified name for the directory should be provided. If you don't, the directory will be searched for in the current directory and it may not be found. Example:

```
#!/usr/bin/python3

import os

# This will delete the "/tmp/testdirectory" directory.

os.rmdir("/tmp/directory")
```

Chapter 12- Tkinter

Python has a number of modules that can be used for GUI (Graphical User Interface) development. An example of such modules is the *tkinter*. The module can be used for development of various GUI elements including buttons, textboxes, image buttons etc. If you need to develop a Python application that will provide its users with a graphical user interface, this is the bets module for you to use.

To create a GUI with the TKInter module, follow these steps:

1. Begin by importing the TKinter module. This is done using the import keyword.
2. Create the main window that will house the GUI elements.
3. Add the GUI elements including buttons, textboxes etc.
4. Add the main event loop to respond to user actions on the interface.

You should note that the module was referred to as *TKinter* in Python 2.7. In Python 3, the name of the module has changed to tkinter. As we had stated earlier, Python is a case sensitive coding language, hence the two are different and failure to adhere to this may generate errors.

The version of Python you are running your code in will determine the output. The names of modules have changed significantly. In my case, I am using Python version 3.5.0.

The following code demonstrates how to create the main window where to add your GUI components in Python:

```
#!/usr/bin/python3
```

```
import tkinter
```

window = tkinter.Tk()

Code for the widgets should be added here

window.mainloop()

Run the code and it will give a window as shown below:

We began by importing the *tkinter* module via the import keyword. If we needed to import everything provided by the module, we would have used this command:

from tkinter import *

The window has then been created by calling the TK() method from the tkinter module as shown below:

window = tkinter.Tk()

As usual, the module name comes first, then the name of the method we are calling.

TKinter Buttons

Buttons can be used for displaying text to text. When a user clicks a button, it should do something. The following code demonstrates how one can create a button with tkinter:

```
#!/usr/bin/python3

from tkinter import *

from tkinter import messagebox

window = Tk()

window.geometry("200x200")

def sayHello():
    msg=messagebox.showinfo( "Hi!", "Thank you for clicking the button")

myButton = Button(window, text ="Hello, Click Me!", command = sayHello)

myButton.place(x=50,y=50)

window.mainloop()
```

In above code, we have imported the *tkinter* module. We also need to use the *messagebox* attribute from the module, hence we also imported it. The window was created on which the button will be added. The *geometry()* method helps us set the size of the window. A function named *sayHello()* has been defined, and this function is to show a message box with some text on it. The *Button()* class helps us create a button named *myButton*. The parameters for the class include the window on which the button is to be displayed, which is *window*, the text to be displayed on the button which is a string and the command, which defines the method to be called once the button is clicked. This has been set to *sayHello()* method.

Running the code gives the following window:

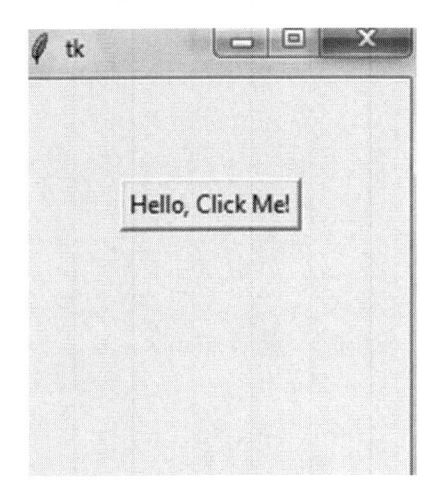

Click the button just as instructed. You will get the following message box:

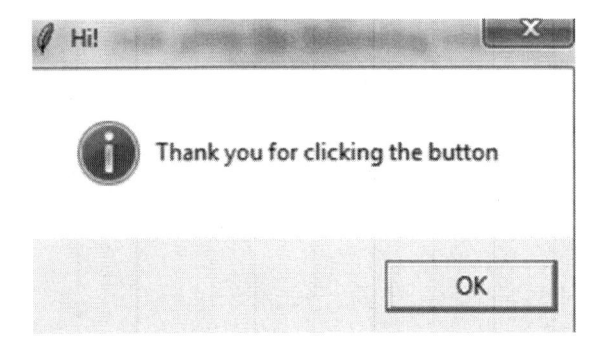

TKinter MenuButtons

In this app, we will demonstrate how one can create a MenuButton and Menus in Python. A menubutton is a drop-down menu which will stay on

130

the screen all the time. Each menubutton has a Menu widget for displaying the choices of the choices for the menubutton once the user has clicked on it.

Example:

```
# !/usr/bin/python3
from tkinter import *
import tkinter
window = Tk()
window.geometry("100x100")
menub= Menubutton ( window, text="File", relief=RAISED )
menub.grid()
menub.menu  =  Menu ( menub, tearoff = 0 )
menub["menu"]  =  menub.menu


newVar  = IntVar()
openVar = IntVar()
menub.menu.add_checkbutton ( label="New",
            variable=newVar )
menub.menu.add_checkbutton ( label="Open",
            variable=openVar )
menub.pack()
```

window.mainloop()

When executed, the code gives the following window:

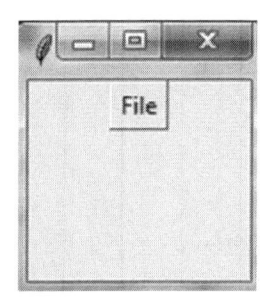

Click on the button written "File" and observe what will happen. A dropdown which shows the options for both "New" and "Open" will show up. If you click on any of these two, it will be checked.

To create a menubutton in Python, we use the "Menubutton()" function and we pass the necessary parameters to it. The "text" parameter specifies the text which will appear on the button. The submenus have to be given a variable name as well as the label which will be visible.

In most applications, menus are very common. They help us to select an option from a list of options. They also help us to save on the space which is provided on the window screen.

In a practical application, menus are very important. The following example demonstrates how these may be used:

!/usr/bin/python3

from tkinter import *

def function1():

  window = Toplevel(root)

```
bt = Button(window, text="Hello!")
bt.pack()


root = Tk()
menub = Menu(root)
mymenu = Menu(menub, tearoff=0)
mymenu.add_command(label="New", command=function1)
mymenu.add_command(label="Open", command=function1)
mymenu.add_command(label="Save", command=function1)
mymenu.add_command(label="Save As...", command=function1)
mymenu.add_command(label="Exit", command=function1)
mymenu.add_separator()
mymenu.add_command(label="Exit", command=root.quit)
menub.add_cascade(label="File", menu=mymenu)
edit = Menu(menub, tearoff=0)
edit.add_command(label="Undo", command=function1)
edit.add_separator()
edit.add_command(label="Copy", command=function1)
edit.add_command(label="Paste", command=function1)
edit.add_command(label="cut", command=function1)
```

edit.add_command(label="Delete", command=function1)

edit.add_command(label="Select All", command=function1)

menub.add_cascade(label="Edit", menu=edit)

help = Menu(menub, tearoff=0)

help.add_command(label="Get Help", command=function1)

help.add_command(label="About this", command=function1)

menub.add_cascade(label="Help", menu=help)

root.config(menu=menub)

root.mainloop()

The code gives the following result once executed:

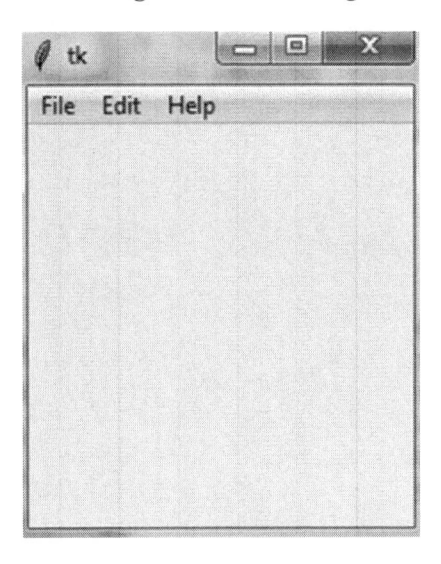

The menus in this case are the File, Edit and Help. Once you click on each of them, you will see that they have a number of options. Note the use of the "add_separator()" method which is used to separate the menus. Click on the File menu and then click on any of the available options. You will see the following window popup:

In this case, we have used the menu to trigger an event. The button shown above with the "Hello" text as created within the function named "function1()". For these menu options, the command has been set to this function. When any of the menus is clicked, it will call the function, which in turn renders the button for us! That is how you can use a button or a menu to trigger an event in Python.

Canvas

In this example, we will demonstrate how you can use a canvas for drawing simple and complex layouts such as pictures. A canvas can accommodate text, widgets, graphics or frames. Here is the code to help you create a canvas:

!/usr/bin/python3

from tkinter import *

window = Tk()

window.geometry("300x300")

cv = Canvas(window, bg="grey", height=220, width=220)

crd = 10, 50, 240, 210

arc = cv.create_arc(crd, start=10, extent=200, fill="red")

line = cv.create_line(20,20,200,200,fill='blue')

cv.pack()

window.mainloop()

The code generates the following once executed:

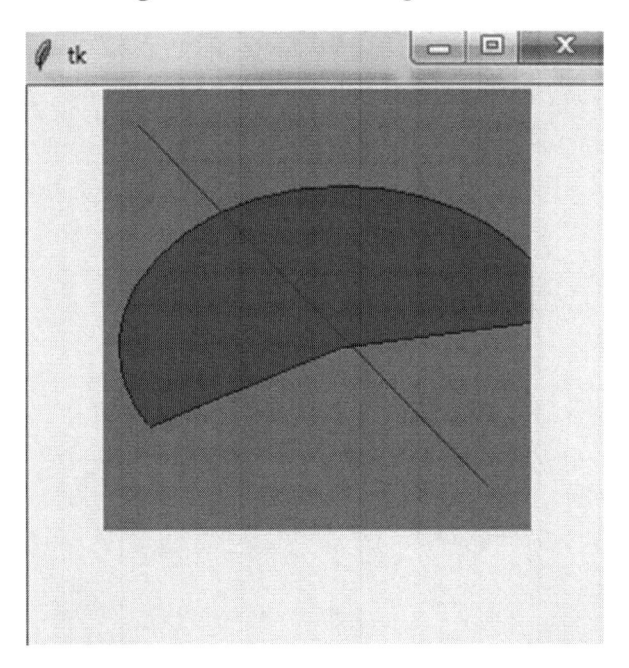

We began by importing our necessary module, the tkinter. Note that the name for this is case sensitive, and that is how we write it in Python 3. We have then used the "Canvas()" function so as to create our canvas. The background color of this canvas has been set to grey, and we have set its height as well as the width.

Our aim is to draw and arch and a line. The arc has to be filled with red color. First, we have used the "crd" variable so as to set the coordinates for this arc. The creation of the arc has been done in the line "arc = cv.create_arc(crd, start=10, extent=200, fill="red")". The use of "cv" means that the arch is being added to the canvas which we created earlier on. The arc should start at 10 and then extend up to 200. Note that to create an arc in Python, we use the "create_arc()" method. We have then added a line to the canvas by use of the "create_line()" method.

Lastly, we have used the "cv.pack()" method so as to render our canvas together with its components, which include the arc and the line.

Slider

It is possible for us to create a graphical slider in Python and allow us to choose some values from it. To do this, we just have to implement a scale widget.

Example:

```
#!/usr/bin/python3

from tkinter import *


def scaleFunction():
    sel = "Value = " + str(x.get())
    lab.config(text = sel)

window = Tk()

x = DoubleVar()

scale = Scale( window, variable = x )

scale.pack(anchor=CENTER)
```

button1 = Button(window, text="See the Value", command=scaleFunction)

button1.pack(anchor=CENTER)

lab = Label(window)

lab.pack()

window.mainloop()

The code results into this when executed:

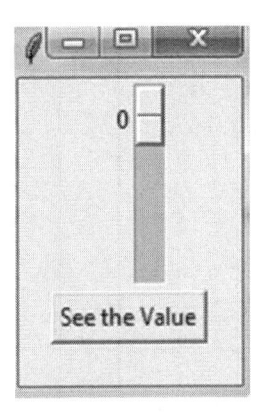

Scroll downwards and you will see that the value of the scale will be shown as you scroll downwards and even upwards. The above figure shows the scale at scale 0. Scroll downwards then click the button written "See the Value".

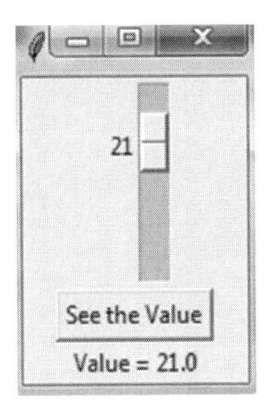

As shown above, the above gets the value of the scale and shows it at the bottom. This shows that we are able to scroll along the scale and then get any value depending on where we are on the scale. We used the "Scale()" function so as to create the scale. The button has been connected to the "scaleFunction()" method, and that is why it is able to get the value which has been selected on the scale.

TKinter Label

This an element that provides you with a box where you can place some text. They are good for labelling. Unlike buttons, labels are not used for responding to user events. However, you can update the text at anytime that you need. The creation of a label involves calling the *Label()* method which takes this syntax:

l = Label (master, option,...)

The master in this case is the parent window on which the label is to be added. The other arguments to the functions are the options for your label, which can act as key value pairs separated by commas. Example:

#!/usr/bin/python3

```
from tkinter import *

window = Tk()

str = StringVar()

lab = Label( window, textvariable = str, relief = RAISED )

str.set("Hi, this is a label")

lab.pack()

window.mainloop()
```

When executed, the code gives the following window:

We created a window and we gave it the name *window*. The label was then created by calling the *Label()* method to which we passed a number of parameters. The string to be added on the label was given the name *str* and its value has also been specified. That is how we got the above label.

TKinter Checkbutton

This is a type of button used when there is a need to display a number of options to the user. The options are represented in the form of toggle button. The user is allowed to select one or more options from the available ones by clicking on the buttons next to the options. One is also allowed to display images in place of text.

The checkbutton is created by calling the *Checkbutton()* method which takes this syntax:

ch = Checkbutton (master, option,..)

The master represents the name of the window on which you need to add the checkbutton. The options for the checkbutton are numerous, and these are determined by the way you need your checkbutton to appear.

Example:

```
# !/usr/bin/python3
from tkinter import *
import tkinter
window = Tk()
Option1 = IntVar()
Option2 = IntVar()
Ch1 = Checkbutton(window, text = "Male", variable = Option1, \
        onvalue = 1, offvalue = 0, height=5, \
        width = 20, )
Ch2 = Checkbutton(window, text = "Female", variable = Option2, \
        onvalue = 1, offvalue = 0, height=5, \
        width = 20)
Ch1.pack()
Ch2.pack()
window.mainloop()
```

When executed, the code will result into the following window:

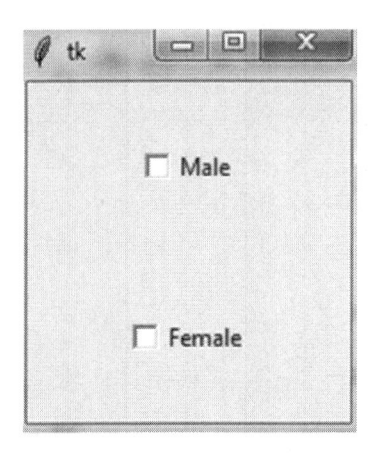

[Screenshot of a tk window showing two checkboxes labeled "Male" and "Female"]

The first checkbutton has been identified as Ch1 while the second one has been identified as Ch2. The string to be added next to each checkbutton has been added with the attribute *text*. The *offvalue* determined the value of the checkbutton when it is deactivated or unchecked, while the onvalue determines the value of the checkbutton when it is activated or checked. The height and the width of the checkboxes have also been specified.

TKinter Radiobutton

With this type of button, the user is provided with a number of buttons to select from, and the user is only allowed to select one of them. This is not the case with the checkbutton as the checkbox allows one to select more than one options. For such a functionality to be achieved with radiobuttons, all radiobuttons in the same group are associated with the same variable. Each button should also be a symbol of a single value.

The tab key helps us switch from one radiobutton to another.

To create a Radiobutton, we must call the *Radiobutton()* method which takes a simple syntax as shown below:

radio = **Radiobutton (master, option,...)**

The master is the first argument which denotes the parent window for the radiobutton. This is then followed by a number of other options for the button. Example:

```
#!/usr/bin/python3

from tkinter import *

def optionSelection():
    option = "You must select an option " + str(x.get())
    lab.config(text = option)

window = Tk()

x = IntVar()

Opt1 = Radiobutton(window, text = "First Option", variable = x, value = 1,
        command = optionSelection)
Opt1.pack( anchor = W )


Opt2 = Radiobutton(window, text = "Second Option", variable = x, value = 2,
        command = optionSelection)
Opt2.pack( anchor = W )


Opt3 = Radiobutton(window, text = "Third Option", variable = x, value = 3,
```

```
              command = optionSelection)

Opt3.pack( anchor = W)

lab = Label(window)

lab.pack()

window.mainloop()
```

When executed, the above code generates this window:

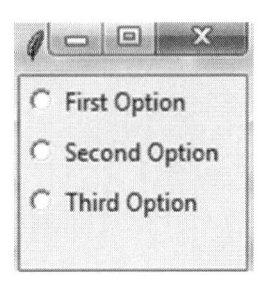

○ First Option
○ Second Option
○ Third Option

Click or activate or select ay of the options and see what happens:

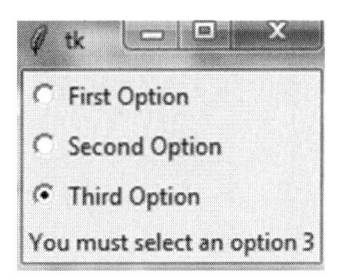

○ First Option
○ Second Option
⦿ Third Option
You must select an option 3

Some text is displayed at the bottom of the window. Notice that we defined only a single variable named x. This variable has been associated with every option via the *variable* method. We also defined a method named *optionSelection*. This has also been associated with every option in the radiobutton group. The method will be called every time you select any of the radiobuttons. This also shows that the *Radiobutton()* method takes the *command* argument, which helps you specify what next after the user has selected the radiobutton. The text displayed after selecting the radiobuttons is shown on a label. You also notice that the interpreter is capable of knowing the option that you have selected, whether 1, 2 or 3.

Chapter 13- Python Operators

Python operators help us manipulate value of operands in operations. Example:

10 * 34 = 340

In the above example, the values 10 and 34 are known as operands, while * is known as the operator. Python supports different types of operators.

Arithmetic Operators

These are the operators used for performing the basic mathematical operations. They include multiplication (*), addition(+), subtraction (-), division (/), modulus (%) and others. Example:

#!/usr/bin/python3

n1 = 6

n2 = 5

n3 = 0

n3 = n1 + n2

print("The value of sum is: ", n3)

n3 = n1 - n2

print("The result of subtraction is: ", n3)

n3 = n1 * n2

print("The result of multiplication is:", n3)

n3 = n1 / n2

print ("The result of division is: ", n3)

n3 = n1 % n2

print ("The remainder after division is: ", n3)

n1 = 2

n2 = 3

n3 = n1**n2

print ("The exponential value is: ", n3)

n1 = 20

n2 = 4

n3 = n1//n2

print ("The result of floor division is: ", n3)

The code prints the followig when executed:

```
The value of sum is:    11
The result of subtraction is:   1
The result of multiplication is: 30
The result of division is:   1.2
The remainder after division is:   1
The exponential value is:   8
The result of floor division is:   5
```

That is how the arithmetic operations work in Python. The modulus operator (%) returns the remainder after a division has been done. In our case, we are dividing 6 by 5, and the remainder is 1.

Comparison Operators

These operators are used for comparing the values of operands and identify the relationship between them. They include the equal to (==), not equal to (!=), less than (<), greater than (>), greater than or equal to (>=) and less than or equal to (<=).

Example:

#!/usr/bin/python3

n1 = 6

n2 = 5

if (n1 == n2):

   print ("The two numbers have equal values")

else:

   print ("The two numbers are not equal in value")

if (n1 != n2):

   print ("The two numbers are not equal in value")

else:

   print ("The two numbers are equal in value")

if (n1 < n2):

   print ("n1 is less than n2")

else:

```
    print ("n1 is not less than n2")
if (n1 > n2 ):
    print ("n1 is greater than n2")
else:
    print ("n1 is not greater than n2")
n1,n2=n2,n1 #the values of n1 and n2 will be swapped. n1=5, n2=6
if ( n1 <= n2 ):
    print ("n1 is either less than or equal to n2")
else:
    print ("n1 is neither less than nor equal to n2")
if ( n2 >= n1 ):
    print ("n2 is either greater than or equal to n1")
else:
    print ("n2 is neither greater than nor equal to n1")
```

The code will print the following:

```
The two numbers are not equal in value
The two numbers are not equal in value
n1 is not less than n2
n1 is greater than n2
n1 is either less than or equal to  n2
n2 is either greater than or equal to n1
```

The value of n1 is 6, while that of n2 is 5. The use of the equal to (==) operator on the two operands will return a false as the two operands are not equal. This will lead the execution of the "else" part. The operator not equal to (!=) will return a true as the values of the two operands are not equal. The only logic which might seem complex in this case is the swapping of the values. The value of n1, which is 6 becomes 5, while that of n2 becomes 6. The statements which are below this swapping statement will then operate with these two new values.

Assignment Operators

These operators the combination of the assignment operator (=) with the other operators. A good example of an assignment operator is "+=". The expression p+=q means "p=p + q". The expression "p/=q" means that "p=p / q". The assignment operators involve combining the assignment operator with the rest of the other operators. Example:

#!/usr/bin/python3

n1 = 6

n2 = 5

n3 = 0

n3 = n1 + n2

print ("The value of n3 is: ", n3)

n3 += n1

print ("The value of n3 is: ", n3)

n3 *= n1

print ("The value of n3 is: ", n3)

n3 /= n1

print ("The value of n3 ", n3)

n3 = 2

n3 %= n1

print ("The value of n3 is: ", n3)

n3 **= n1

print ("The value of n3 is: ", n3)

n3 //= n1

print ("The value of n3 is: ", n3)

The code will print the following when executed:

```
The value of n3 is:    11
The value of n3 is:    17
The value of n3 is:    102
The value of n3      17.0
The value of n3 is:    2
The value of n3 is:    64
The value of n3 is:    10
```

The statement "n3 = n1 + n2" is very straight forward as we are just adding the value of n1 to that of n2. In the expression "n3 += n1", we are adding the value of n3 to that of n1 and then assign the result to n3. However, note that in the previous statement, the value of n3 changed to "11" after adding n1 to n2. So we have 11+6, which gives 17. After that, the new value of the variable n3 will be 17. The expression "n3 *= n1" means "n3= n3 * n1". This will be 17 * 6, and the result will be 102. That is how these operators work in Python!

Membership Operators

These are the operators which are used for testing membership in a certain sequence of elements. The sequence of elements can be a string, a list or a tuple. The two membership operators include "in" and "not in".

The "in" operator returns true if the value you specify is found in the sequence. The operator "not in" will evaluate to a true if the specified element is not found in the sequence. Example:

```
#!/usr/bin/python3

n1 = 7

n2 = 21

ls = [10, 20, 30, 40, 50 ]

if ( n1 in ls ):

   print ("n1 was found in the list")

else:

   print ("n1 was not found in the list")

if ( n2 not in ls ):

   print ("n2 was not found in the list")

else:

   print ("n2 was found in the list")

n3=n2/n1

if ( n3 in ls ):
```

print ("n1 was found in the list")

else:

print ("n1 was not found in the list")

The code will print the following once executed:

```
n1 was not found in the list
n2 was not found in the list
n1 was not found in the list
```

The value of num1 is 7. This is not part of our list, and that is why the use of the "in" operator returns a false. This causes the "else" part to be executed. The value of n2 is 21. This is not in the list. This expression returns a true, and the first part below the expression is executed. 21 divide by 7 is 3. This value is not in the list. The use of the last "in" operator evaluates to a false, and that is why the "else" part below it is executed.

Identity Operators

These operators are used to compare the values of two memory locations. Python has a method named "id()" that returns the unique identifier of the object. Python has two identity operators:

is- this operator evaluates to a true in case the variables used on either sides of the operator are pointing to a similar object. It evaluates to false otherwise.

is not- this operator evaluates to a false if the variables on either sides of the operator are pointing to a similar object, and true otherwise.

Example:

#!/usr/bin/python3

```python
n1 = 45
n2 = 45
print ('The initial values are','n1=',n1,':',id(n1), 'n2=',n2,':',id(n2))
if ( n1 is n2 ):
   print ("1. n1 and n2 share an identity")
else:
   print ("2. n1 and n2 do not share identity")
if ( id(n1) == id(n2) ):
   print ("3. n1 and n2 share an identity")
else:
   print ("4. n1 and n2 do not share identity")
n2 = 100
print ('The variable values are','n1=',n1,':',id(n1), 'n2=',n2,':',id(n2))
if ( n1 is not n2 ):
   print ("5. n1 and n2 do not share identity")
else:
   print ("6. n1 and n2 share identity")
```

The code will print the following once executed:

```
The initial values are n1= 45 : 1730008176 n2= 45 : 1730008176
1. n1 and n2 share an identity
3. n1 and n2 share an identity
The variable values are n1= 45 : 1730008176 n2= 100 : 1730009936
5. n1 and n2 do not share identity
```

Note that I have numbered some of the print statements so that it may be easy to differentiate them. In the first instance, the values of variables n1 and n2 are equal. The first statement of the output shows the respective values for the variables together with their unique identifier. Note that the identifier has been obtained by use if the id() Python method, and the name of the variable has been passed inside the function as the argument. The expression "if (n1 is n2):" will evaluate to a true since the values of the two variables are equal, or they are pointing to a similar object. This is why the print statement labeled 1 was executed!

You must also have noticed that the unique identifiers of the two variables are equal. In the expression "if (id(n1) == id(n2)):", we are testing whether the values of the identifiers for the two variables are the same. This evaluates to a true, hence the print statement labeled 3 has been executed!

The expression "n2 = 100" changes the value of variable n2 from 45 to 100. At this point, the values of the two variables will not be equal. This is because n1 has a value of 45, while n2 has a value of 100. This is clearly in the next print statement which shows the values of the variables together with their corresponding ids. You must also have noticed that the ids of the two variables are not equal at this point.

The expression "if (n1 is not n2):" evaluates to a true, hence the print statement labeled 5 was executed. If we test to check whether the values of the ids for the two variables are equal, you will notice that they are not equal.

Chapter 14- Accessing MySQL Databases

It is possible for you to access a MySQL database from your Python code. For the of SQLite, Pytho has an in-built support for it. You can use an interface named PyMySQL to connect to a MySQL database from Python. To know whether this package has been installed on your computer, you have to open the Python terminal then run the following command:

import **PyMySQL**

The command simply tries to import the module. If it is not found, you will get an error.

```
>>> import PyMySQL
Traceback (most recent call last):
  File "<stdin>", line 1, in <module>
ImportError: No module named 'PyMySQL'
>>>
```

The above figure shows that the module is not installed. I must install it. Python comes with a module named pip that helps you install other Python modules and packages. You can use it to install PyMySQL by running the following command from the terminal of your operating system:

pip install **PyMySQL**

The command will run and PyMySQL will be installed on your system:

```
C:\Users\admin>pip install PyMySQL
Collecting PyMySQL
  Downloading https://files.pythonhosted.org/packages/32/e8/222d9e1c7821f935d6db
a8d4c60b9985124149b35a9f93a84f0b98afc219/PyMySQL-0.8.1-py2.py3-none-any.whl (81k
B)
    100% |################################| 81kB 178kB/s
Installing collected packages: PyMySQL
Successfully installed PyMySQL-0.8.1
You are using pip version 9.0.1, however version 10.0.1 is available.
You should consider upgrading via the 'python -m pip install --upgrade pip' comm
and.

C:\Users\admin>
```

The above figure shows the installation was successful.

Now that the module has been installed, you can use I to connect to the MySQL server. However, ensure that you have created a database that you need to connect to.

In my case, I will be connecting to a database named *school*. The database has a table named *student* with four columns namely name, admission, course and age.

```
mysql> desc student;
+-----------+-------------+------+-----+---------+-------+
| Field     | Type        | Null | Key | Default | Extra |
+-----------+-------------+------+-----+---------+-------+
| name      | varchar(30) | YES  |     | NULL    |       |
| admission | varchar(30) | YES  |     | NULL    |       |
| course    | varchar(30) | YES  |     | NULL    |       |
| age       | int(11)     | YES  |     | NULL    |       |
+-----------+-------------+------+-----+---------+-------+
4 rows in set (0.00 sec)

mysql>
```

The following Python code can help us establish a connection to the School database:

#!/usr/bin/python3

import PyMySQL

Opening the database connection

```python
database = PyMySQL.connect("localhost","root","","SCHOOL" )

# create the cursor object by calling cursor() method

cursor = database.cursor()

# execute a SQL query by calling execute() method.

cursor.execute("SELECT VERSION()")

# Fetch single row by calling fetchone() method.

dt = cursor.fetchone()

print ("Database version : %s " % dt)

# disconnect from the server

database.close()
```

The code will print the version of MySQL that you have installed on your system. If you are using either WampServer or XAMPP, ensure that they are up and running before running the above code. Consider the following line extracted from the above code:

```python
database = PyMySQL.connect("localhost","root","","SCHOOL" )
```

We have simply invoked the *connect()* method which is contained in the *PyMySQL* module. The method takes a number of parameters. The first one is the computer on which the MySQL server is running on, which in our case is the *localhost*. The next parameter is the *username* for the MySQL account we need to use for connection. We are logging in as the *root* user. The next parameter is the password of the username you have provided. In my case, the password for root user is blank, hence we enter nothing there. The last parameter is the name of the MySQL database we need to connect to. This has been specified as the *SCHOOL* database.

Whenever we need to run a SQL statement, we call the *execute()* method as we have done above. To close a connection to a MySQL database, we call the *close()* method as done above.

Creating a Table

Now that we have established a connection to the database, we can create tables or even insert records into the existing tables. We will create a table named *Workers*. This can be done via the *execute* method:

#!/usr/bin/python3

import PyMySQL

Opening the database connection

database = PyMySQL.connect("localhost","root","","SCHOOL")

preparing a cursor object by calling cursor() method

cursor = database.cursor()

Drop the table if it exists by calling execute() method.

cursor.execute("DROP TABLE IF EXISTS WORKERS")

Creating a table as required

statement = """CREATE TABLE WORKERS (

  FIRST_NAME CHAR(30) NOT NULL,

  LAST_NAME CHAR(30),

  AGE INT,

  GENDER CHAR(1),

SALARY FLOAT)"""

cursor.execute(statement)

disconnecting from the server

database.close()

After that, confirm from the MySQL terminal to check whether the table has been created.

Inserting Data

Let us insert some data into the WORKERS table that we have just created:

```
#!/usr/bin/python3

import PyMySQL

# Opening the database connection

database = PyMySQL.connect("localhost","root","","SCHOOL")

# preparing the cursor object by calling cursor () method

cursor = database.cursor()

# Preparing a SQL statement to INSERT record into our table.

statement = """INSERT INTO WORKERS (FIRST_NAME,
   LAST_NAME, AGE, GENDER, SALARY)
   VALUES ('Nicholas', 'Samuel', 26, 'M', 5000)"""

try:
   # Execute the above SQL command
```

```
    cursor.execute(statement)

    # Commit the changes made to the database

    database.commit()
except:
    # Rollback in case an error occurs

    database.rollback()
# disconnecting from the server

database.close()
```

The record will be inserted into your table. If you need the code to generate the queries dynamically, you can modify it to the following:

```
#!/usr/bin/python3

import PyMySQL

# Opening the database connection

database = PyMySQL.connect("localhost","root","","SCHOOL" )

# preparing the cursor object by calling cursor() method

cursor = database.cursor()

# Preparing a SQL statement to INSERT record into our table.

statement = "INSERT INTO WORKERS(FIRST_NAME, \
   LAST_NAME, AGE, GENDER, SALARY) \
   VALUES ('%s', '%s', '%d', '%c', '%d' )" % \
   ('Nicholas', 'Samuel', 26, 'M', 5000)
```

try:

   # Execute the above SQL command

   cursor.execute(statement)

   # Commit the changes made to the database

   database.commit()

except:

   # Rollback in case an error occurs

   database.rollback()

disconnecting from the server

database.close()

After that, you will only have to change the values for each worker and their details will be inserted into the database table.

Conclusion

Python is a good programming language that can be used for development of various types of applications. It is an object-oriented programming language which supports the use of the object-oriented programming concepts. Most of the aspects of Python have been discussed in this book, and my hope is that you are now familiar with them. That is all what it takes for one to become a Python geek. The language is interpreted rather than being compiled. If you are a beginner to programming, Python is a good language to begin with. It will help you form a solid background that make it easy for you to learn any other type of programming language. This is because Python comes with numerous constructs that are easy for you to learn. These constructs are also closely related to the constructs used in other programming languages.

Thank you!

Thank you for buying this book! It is intended to help you understanding Python Programming. If you enjoyed this book and felt that it added value to your life, I ask that you please take the time to review it.

Your honest feedback would be greatly appreciated. It really does make a difference.

THE BEST WAY TO THANK AN AUTHOR IS TO WRITE REVIEW BY CLIKING <u>HERE</u>

Printed in Great Britain
by Amazon